D0904154

The World of Sound

THE SCIENCE OF ACOUSTICS

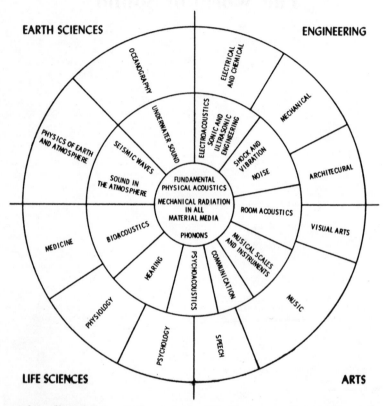

This illustration of the relationship between the science of Acoustics and the various arts and sciences was designed by Dr. R. Bruce Lindsay and was included in a report to the National Science Foundation on a Conference on Education in *Acoustics* by Dr. Lindsay, published in the *Journal of the Acoustical Society of America*, 36, 2241–2243 (1964). Reproduced by permission of Dr. Lindsay and the American Institute of Physics.

The World of Sound

Vernon M. Albers

South Brunswick and New York:
A. S. Barnes and Company
London: Thomas Yoseloff Ltd

A. S. Barnes and Co., Inc.
Cranbury, New Jersey 08512

Thomas Yoseloff Ltd
108 New Bond Street
London W1Y OQX, England

BY THE SAME AUTHOR

Amateur Furniture Construction
Repair and Reupholstering of Old Furniture
Underwater Acoustics Handbook
Underwater Acoustics Handbook II
Underwater Acoustics Instrumentation

ISBN 0-498-07676-8

Printed in the United States of America

Contents

Introduction

Chapter	1. *The Nature of Sound*	17
	1.1. Vibration.	17
	1.2. Forced Vibration.	23
	1.3. Waves.	24
	1.4. Sound Intensity, Sound Pressure and Particle Velocity.	34
	1.5. Sound.	36
Chapter	2. *Units and Reference Quantities Used in Measuring Sound*	38
	2.1. The Decibel Scale for Measuring Sound Levels.	38
	2.2. Use of dB Scales in Measurements.	43
	Appendix: The Decibel Scale	46
Chapter	3. *Speech and Hearing*	50
	3.1. Speech.	50
	3.2. Hearing.	52
Chapter	4. *Sound Propagation*	60
	4.1. Divergence.	60
	4.2. Reflection.	61
	4.3. Refraction.	65
	4.4. Attenuation.	68
	Appendix: Snell's Law	69
Chapter	5. *Noise*	72
Chapter	6. *Transducers*	77
	6.1. Introduction.	77
	6.2. Transducer Elements.	78

6.3. Microphones.   80
6.4. Methods of Incorporating Transducing
Elements in Transducers.   83
6.5. Transducer Arrays and Directivity.   92
6.6. Non Reciprocal Transducers.   94

Chapter   7. *Sound Measurement and Analysis*   95
7.1. Introduction.   95
7.2. Measurement of Sound Pressure Levels.   96
7.3. The Sound Level Meter.   98
7.4. Measurement of Sound Pressure Levels as
a Function of Frequency.   100
7.5. Narrow Band Spectrum Analysis.   102
7.6. The Sound Spectrograph.   106

Chapter   8. *Audiometry and Noise Hazard*   112
8.1. Introduction.   112
8.2. Masking.   113
8.3. Hearing Threshold Shifts Due to Noise.   114
8.4. Acceptable Noise Exposure.   117
8.5. Industrial Noise Control.   120
8.6. Personal Protection.   120
8.7. Legal Aspects of Noise Hazard.   122
8.8. Community Noise.   123

Chapter   9. *Sound Recording and Reproduction*   125
9.1. Early Methods of Sound Recording and
Reproduction.   125
9.2. Modern Phonograph Disc Recording and
Reproduction.   128
9.3. Optical Recording and Reproduction.   133
9.4. Magnetic Recording.   135
9.5. Distortion.   139

Chapter 10. *The Physics of Music*   144
10.1. Introduction.   144
10.2. Pitch.   145
10.3. Quality.   148
10.4. Musical Intervals.   150
10.5. Musical Scales.   153

Contents

Chapter 11. Musical Instruments   158
    11.1. Vibrating Bars.   158
    11.2. Vibrating Strings.   162
    11.3. Vibrating Air Columns.   169
    11.4. The Pipe Organ.   175
    11.5. Other Vibrating Air Column Musical
          Instruments.   175
    11.6. Vibration of Stretched Membranes.   176
    11.7. Cymbals and Bells.   178
    11.8. Electric Instruments.   179

Chapter 12. Room Acoustics   180
    12.1. Introduction.   180
    12.2. Sound Insulation.   181
    12.3. Reverberation.   182
    12.4. Echoes.   185
    12.5. Sound Reinforcement.   188

Chapter 13. Underwater Sound   192
    13.1. Introduction.   192
    13.2. Generation and Detection of Underwater
          Sound.   193
    13.3. Propagation of Sound in the Ocean.   198
    13.4. Studies of Marine Animals.   206
    13.5. Bottom Topography.   208
    13.6. Marine Geology.   211
    13.7. Underwater Communication.   214
    13.8. Miscellaneous Applications of Underwater
          Sound.   215

Chapter 14. Utilization of Sound by Animals   218
    14.1. Introduction.   218
    14.2. Utilization of Sound by Man.   218
    14.3. Use of Sound by Animals as Danger and
          Courting Signals.   220
    14.4. Use of Sound for Navigation and Food
          Hunting.   222

Chapter 15. Careers in Acoustics   224
    15.1. Introduction.   224

The World of Sound

15.2. Physics. 225
15.3. Architectural Engineering. 225
15.4. Fluid Mechanics. 226
15.5. Electrical Engineering. 227
15.6. Mechanical Engineering. 228
15.7. Biology. 229
15.8. Community Planning. 230
15.9. Public Health. 230
15.10. Oceanography. 231
15.11. Marine Geology. 232

Supplementary Reading 233

Index 235

Introduction

Sound is very important in our lives. It is an important means of communication; it can be a source of pleasure as we listen to music and it can be a source of discomfort when we are forced to listen to distracting noises. Few of us have ever experienced a complete absence of sound; in fact, such an absence of sound can be quite disconcerting to us.

The human ears, with their associated analysis system in the brain, is a marvelous system, it is a system capable of functioning with an almost inconceivably low level of sound intensity and one that can function effectively at the sound intensity of a rock 'n' roll orchestra producing more than 10 billion times as great intensity at the ear as the lowest intensity that the ear can detect.

The science of sound is usually termed acoustics. There is scant reference to the science of sound prior to the middle of the seventeenth century. The earlier interest in the subject was primarily concerned with musical sounds and musical instruments, so it is necessary to refer to musical works to trace the earlier growth of the subject.

Our existing system of music is derived from the Greeks. However, the art of music was practiced by Hindus, Egyptians, Chinese and Japanese as early as 4000 B.C.

Pythagoras originated the studies of musical intervals

and ratios; he used a monochord, which consists of a string maintained at a uniform tension passing over two wooden bridges. He determined that under the same tension the ratio of the lengths of the string sounding the fundamental and its octave was 2–1, but it was not until the middle of the seventeenth century that Mersenne and Galileo carried the experiments of Pythagoras a step further and determined the effect of the mass and tension of the string on its frequency of vibration.

We are indebted to the Roman architect Vitruvius for our knowledge of the acoustic characteristics of Greek theatres. He demonstrated his knowledge of the role of interference, echoes, and reverberation on the acoustic characteristics of a theatre in a work published about 20 B.C.

Galileo (1564–1642) laid the foundations of experimental acoustics. He measured the effects of mass and tension on the frequency of vibration of a string independently of the corresponding measurements by Mersenne (1588–1648).

The velocity of sound in air was first measured by Borelli and Viviani of the Accademia del Cimento in Florence. They obtained a value of 1148 ft/sec. Cassini, Römer, Picard and Huygens of the Paris *Academie des Sciences* measured a value of 1142 ft/sec. These measurements were made prior to 1667. In 1808 Biot measured the velocity of sound in an iron pipe, and in 1826 Colladon and Sturm determined the velocity of sound in water in Lake Geneva.

Newton studied the theoretical aspect of sound and derived the formula for the velocity of sound, which states that the velocity is equal to the square root of the elasticity of the medium divided by its density for the velocity of a pulse in an elastic medium. This was published in Book II of his *Principia* in 1687. The velocity of sound in air, which he calculated from this formula, was 968 ft/sec. About 70 years later Lagrange (1736–1813) pointed out that Newton's calculation failed to take into consideration the changes in elasticity of the air due to the temperature changes produced by the propagation of the sound waves. Pierre Laplace modfied Newton's formula to calculate a velocity that agreed with the experimental values.

H. V. Helmholz (1821–1894) and Lord Rayleigh (1842–1919) were the men mainly responsible for the establishment of acoustics as an exact science. Helmholz occupied the chair of Physiology at Königsberg in 1849, of Anatomy at Bonn in 1855 and of Physics at Berlin in 1871. His contribution was primarily physiological. He published his *Sensations of Tone* in 1862 and he developed the theory of summation and difference tones and the theory of resonators. He collaborated with a pianoforte maker both in improving and testing the instruments.

Lord Rayleigh was primarily interested in Physics. His book, *The Theory of Sound*, gave a a survey of the subject up to that time and he developed considerable theory which opened up a vast field for later experimenters. He was also an accomplished experimenter. He developed a delicately suspended disc which tends to set itself perpendicular to the direction of propagation of a wave. By means of this device he was able to measure the absolute intensity of sound.

John Tyndall (1820–1893), who succeeded Michael Faraday as Director of the Royal Institution, carried out many original experiments on singing and sensitive flames. He did much by means of his lectures to popularize an interest in sound.

K. R. Koenig (1839–1901) studied under Helmholz but became an instrument maker and did much research on sound. He developed a tonometric apparatus involving about 600 tuning forks and he developed a "phonautograph," which was the predecessor of Edison's phonograph.

Sir Charles Wheatstone (1802–1875) invented an instrument similar to a stethoscope and originated the term "microphone." He used a rotating mirror to study the motion of vibrating systems.

George S. Ohm (1784–1854) formulated the law that states, "all types of tone quality are due to certain combinations of simple tones whose frequencies are commensurable and a complex tone may be analyzed into a sum of simple tones."

Chladni (1756–1827) used powder figures to study the vibration of plates. His experiments are described in his book *Die Akustik.*

The submarine campaign in the first world war was a factor in the rapid development of the science of acoustics. The development of the vacuum tube also stimulated the demand for improved sound recorders and reproducers.

The increased use of mechanization in modern life has stimulated an interest in noise and noise isolation. The problem of environmental acoustics is assuming increased importance. Noise in factories causes problems of hearing loss. Noisy airplanes make living conditions near airports undesirable and much needs to be done before we will know how to design a good auditorium or concert hall.

Introduction

There are many treatises on sound and acoustics, but these are generally quite mathematical. In 1877 Lord Rayleigh published his treatise on *Theory of Sound,* which treated the mathematical theory of acoustics and much work has been done since that time to further develop a rigorous theory of the various phenomena related to acoustics.

I will, however, attempt to give a description of the generation, propagation, detection and measurement of sound without the use of mathematics. In a few instances, I have included appendices containing simple mathematical treatment of material that is discussed without mathematics in the text.

Although sound is one of the classical divisions of physics, it does not receive the attention in the high school and college physics courses that its importance in our modern civilization warrants. We are learning the importance of excessive noise in our environment and we do not have agreement on how to properly define what is excessive noise and how to measure and control it. We are gradually learning how to record and reproduce music so that it resembles the original music; but we are far from attaining perfection in the process. We have a very inadequate understanding of the design of musical instruments and we are just beginning to learn how animals utilize sound in their lives.

It is hoped that this book will be useful to the many people who must deal with problems in acoustics—particularly in the fields of noise and noise control and audiometry—but who do not have the necessary mathematics and physics background to utilize the other available textbooks.

There are many kinds of careers in acoustics open to

people who are interested, and as yet there are few people being trained in the field. Most of the people working with sound were trained in other areas, such as physics, engineering and biology, but they have been attracted to acoustics because of the many fascinating and challenging problems to be solved. I hope that this book will help to interest young students in the study of sound.

1.

The Nature of Sound

1.1. *Vibration.*

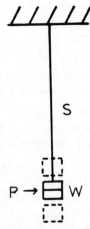

Fig. 1.1. A simple mass-spring vibrator.

Figure 1.1 shows a weight W suspended from a rigid support by means of a rubber band S. If we pull the weight down slightly and then release it, the weight will bob up and down. If we observe the motion of a line P on the weight as it moves up and down, we will find that as it crosses the center of the motion it moves with its highest velocity, and that velocity is equal to zero at each end of the path of motion. The type of motion executed by the

17

mark on the weight is the same as that of a point P on a
wheel rotating with a uniform angular speed on a shaft
S when it is viewed toward the edge of the wheel as indicated in Fig. 1.2. This kind of motion is called *simple
harmonic motion.*

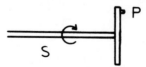

Fig. 1.2.   Illustration of simple harmonic motion.

Because the wheel in Fig. 1.2 is rotating, we can define
the position of the point P in terms of an angle through
which the wheel has turned from some arbitrary starting
position. There are 360 degrees in a complete circle, so
the point P on the wheel in Fig. 1.2 will have traveled
through 360 degrees when it returns to its original position moving in its original direction. The motion repeats
itself every time the wheel makes a complete revolution.
We say, therefore, that the motion of the point P completes one *cycle* each time the wheel makes a complete
revolution and one cycle corresponds to 360 degrees of
rotation of the wheel. If the wheel rotates at a rate of f
revolutions per second, the point P, as viewed from the
edge of the wheel, will move through f cycles per second.
    Since the motion of the weight in Fig. 1.1 is similar to
that of the motion of the projection of the point P on the
wheel, the bobbing weight moves with simple harmonic
motion, and even though no angular motion is involved
its motion can be referred to the corresponding uniform
circular motion in Fig. 1.2, and we can equate one cycle
of the up and down motion of the weight to 360 degrees.

Referring again to Fig. 1.1, we can vary the weight of W and the stiffness of the rubber band S. If the weight is increased, keeping the stiffness of the rubber band constant, the frequency will decrease, but if the weight is held constant and the stiffness of the rubber band is increased, the frequency will increase. You can verify this by determining the frequency of vibration of the system with different weights and different rubber bands. The weight and rubber band combination is a vibrating system. Since practically all of the mass of the system is concentrated in the weight W, and all of the deformation takes place in the rubber band, the mass and elastic portions of the system are obviously separated. This is not true in many vibrating systems. For example, the system illustrated in Fig. 1.3 consists of an elastic bar anchored

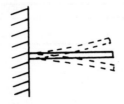

Fig. 1.3.   Vibrating rod fixed at one end.

in a rigid wall. If we strike the end of the bar it will vibrate. The mass of the system is distributed over the length of the bar, and as it vibrates the bar bends over its entire length.

Most vibrating systems can vibrate in several modes. This can be easily illustrated by means of a vibrating string. The string has its mass distributed over its entire length and the elastic restoring force is determined by the tension in the string, which can be controlled by the magnitude of the weight W in Fig. 1.4. The two bridges B_1 and B_2 establish the length of string that can vibrate. If the string is plucked midway between B_1 and B_2, its mode of vibration will be primarily that illustrated in Fig. 1.5. This is called the fundamental mode of vibration. If the string is plucked approximately one fourth of its length from B_1, it will vibrate in the modes illustrated in Fig. 1.6. The fundamental mode can be damped by touching the string at the midpoint and it will then vibrate only in the second mode. The strings in musical instruments are normally struck or plucked near one of the bridges in order to excite higher modes in addition to the fundamental.

Flat plates are capable of vibrating in a large number of modes, some of them quite complex. The vibrations of

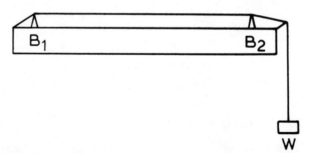

Fig. 1.4. Arrangement for studying vibrating strings.

Fig. 1.5. Fundamental mode of vibration of a string.

Fig. 1.6.   Fundamental and second harmonic modes of vibration of a string.

plates was first studied by Chladni. He clamped circular and square metal plates at a point as illustrated in Fig. 1.7 by means of a large C clamp and excited the plates to vibration by stroking the edge with a violin bow. In order to study the vibration of the plates, he sprinkled powder on the top surfaces. The vibration of the plate causes the powder to move to the areas of no vibration, called the *nodes*. Chladni produced large numbers of powder patterns by damping the vibrations at various points, thereby forcing the plates to vibrate in different modes, producing different nodal patterns.

The back and front of a violin or guitar and the sound-

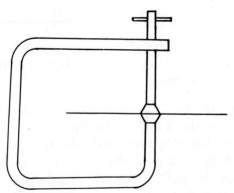

Fig. 1.7.   Arrangement used by Chladni for studying the modes of vibration of plates.

ing board on a piano are examples of vibrating plates that are excited by the strings of the instrument and serve to amplify the sound from the strings. The metal panels on an automobile body tend to vibrate in response to excitation from the engine and road roughness. Since these vibrations are undesirable, the panels on the automobile have various kinds of damping material applied to them in order to suppress these vibrations.

It is possible to excite an air column to vibration. The simplest means of illustrating such vibration is to blow across the neck of a bottle or jug. Although the previous examples were such that the vibration could be observed visually, it is not possible to see the air column vibrate. However the vibration of the air column can be detected by the fact that sound is emitted. By careful control of the method of blowing across the neck of the bottle, it is usually possible to excite more than one mode of vibration as evidenced by the different pitches of sound that can be produced.

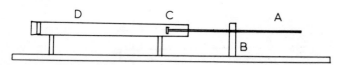

Fig. 1.8.   Kundt's tube method of observation of vibration in an air column.

A method of visualizing the vibration of an air column is provided by the Kundt's tube illustrated in Fig. 1.8. *A* is a ¼-inch diameter brass rod about three feet long clamped at its center by the support *B*. *C* is a piston mounted on the end of the rod. The portion of the rod at *A* can be stroked with a rosin-coated cloth to cause it to vibrate and emit a high-pitched squeal. A glass tube

D about three feet long and about 1½ inches in diameter is supported so that the piston projects a few inches into its open end. If a layer of fine powder is placed along the bottom of the tube, and the tube is slowly adjusted along its supports while the rod is being stroked, a length of air column can be found such that the vibration of the air in the tube will excite the powder in the regions of maximum vibration. The powder will tend to move from the regions of large vibration to the nodes and the powder renders the vibration of the air column visible.

1.2. *Forced Vibration.*

All practical vibrating systems are more or less damped —i.e., when they are set into vibration the amplitude of the vibration will gradually decrease due to friction in the system. If the vibrating system shown in Fig. 1.1 is driven by a source such as a vibrating piston connected to the bottom of the weight by means of a rubber band, and the frequency of the piston is gradually increased from a very low value to a high value, the amplitude of the vibration of the weight W will increase, reaching a maximum value when the frequency is equal to that of the original freely vibrating system. If the driving frequency is increased still further, the amplitude of the vibration will decrease as the frequency is increased. Figure 1.9 shows the plots of the amplitude of vibration as a function of frequency for different degrees of damping. The frequency at which the amplitude of vibration reaches a maximum is the resonant frequency. The two frequencies, f_1 and f_2, are the frequencies above and below the resonant frequency where the amplitude is equal to $1/(2)^{1/2}$ of the maximum amplitude. The frequency band f_2—f_1 is called

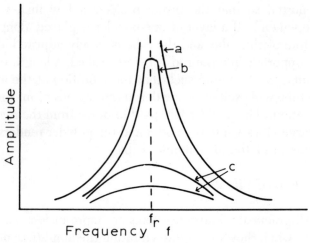

Fig. 1.9. Resonance curves: a no damping, b low damping
        and c high damping.

the band width of the system and $f_r/(f_2-f_1)$ is called the
sharpness of the resonance or the Q of the system.

1.3. Waves.

If you performed the experiments described in the pre-
ceding section, you no doubt observed that you could
hear sound as a result of the vibrations in some instances.
It is always necessary to have some vibrating source if
sound is to be produced. The sound energy is propagated
to our ears through the intervening medium, which is the
air around us. It is possible to demonstrate that air—or
some other material medium—is necessary for the sound
to be propagated by means of the experiment diagrammed
in Fig. 1.10. The bell B in Fig. 1.10 is suspended by a
thread within a bell jar, which can be evacuated by means

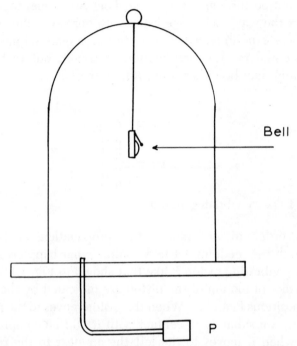

Bell

Fig. 1.10. Arrangement to demonstrate that a material medium is essential for propagation of sound.

of an air pump *P*. If the bell is operated with normal at-
mospheric pressure in the bell jar, it is possible for the
sound of the bell to be heard. If the air pump is started,
we will observe that the ringing of the bell will become
fainter and will finally become inaudible when a high
vacuum is attained within the bell jar. By substituting
other gases for the air in the bell jar, we can demonstrate
that sound will be propagated through them.

    If we attach the bell to the jar by means of a solid sup-
port the sound will be transmitted from the bell jar even

when it is evacuated. We can demonstrate that sound is transmitted through water by striking two stones together under the water while we listen with our ears submerged.

These experiments demonstrate that a material medium is essential for the propagation of sound, but that this medium may be either solid, liquid or gas.

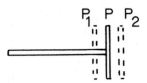

Fig. 1.11.   A vibrating piston.

In order to understand the propagation of sound through the medium, let us visualize what happens if we have a vibrating piston P like that shown in Fig. 1.11. The extremes of motion of the piston are indicated by the dotted positions P_1 and P_2. When the piston moves to the right, the air immediately in contact with it will be compressed and when it moves to the left the air next to the piston will be rarefied. When the air is compressed the molecules of the air will move to the right and when it is rarefied the molecules will move toward the left. The motion of the molecules of the air can be simulated by the arrangement shown in Fig. 1.12. As the piston moves toward the right, it will compress the spring number 1 causing pendulum 1 to start to move to the right. As this pendulum starts to move to the right it will compress spring number 2 and the process will repeat itself down the line of springs and pendulums. When the piston moves to the left, spring number 1 will be stretched causing pendulum number 1 to start to move to the left, which

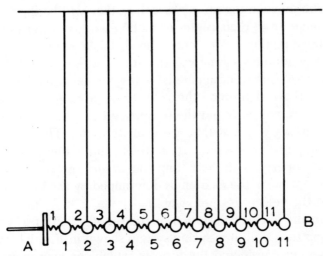

Fig. 1.12.   An arrangement for demonstrating how sound is propagated through an elastic medium.

in turn will stretch spring number 2; this motion will progress down the series of springs and pendulums. If we concentrate our attention on one of the pendulums we will observe that it moves back and forth along the line AB. In order for a pendulum to move, energy must be imimparted to it. That energy is passed from the piston to the first pendulum by way of the first spring, after which it is passed, in turn, from one pendulum to the next. This pendulum-spring system, therefore, serves as a medium that can transfer the energy of the vibrating piston from A to B. This means of propagation of energy is called a compressional wave. Sound is propagated through solid, liquid and gaseous media by means of compressional waves.

     If we watch the motion of the pendulums in Fig. 1.12

while the piston is vibrating, we can select two pendulums which move in the same *phase*, i. e., they will be the same distance from their center of motion while moving in the same direction. We define the distance between these two pendulums as one wavelength. The wavelength of a wave is usually designated by the Greek letter lambda (λ). If we determine the velocity of propagation of the wave from A to B and the frequency f of vibration of the piston, we will find that the velocity $c=f\lambda$. The velocity of propagation is a property of the medium, the frequency is a property of the vibrating source and the wavelength of the wave in the medium is determined by the frequency and the velocity of propagation.

In the above description of the propagation of a wave along the pendulum-spring system shown in Fig. 1.12, we have not considered the form in which the energy exists in the medium as it is being propagated. Also, we have not indicated the possibility of any loss of energy in the process of the propagation and we have not considered what happens when the wave reaches the end of the row of pendulums and springs.

If we compress or stretch a spring, it is necessary to do work on it, but the energy expended is stored in the spring and can be recovered when the spring again returns to its original length. We call this form of energy *potential energy*. This is obvious when we deal with a spring, but the same process operates when we compress or expand an elastic gas liquid or solid. If we start an object in motion, it is necessary to do work on it. The amount of work necessary to initiate a given velocity of motion depends on the mass of the object. This can be illustrated by comparing the amount of work necessary to start a handball down the lane of a bowling alley to the amount of work required to start a bowling ball rolling with the same speed. The energy imparted to start an object in motion can be recovered in the process of stopping the motion. The bowling ball will be able to do a great deal more work on the pins than the handball traveling at the same speed. The energy of motion possessed by a moving object is called *kinetic energy*. The particles or molecules in a medium have mass, so that when they are set in motion they will possess kinetic energy.

Referring again to Fig. 1.12, we can see that the propagation of the wave from A to B is a process of converting potential energy in spring 1 to kinetic energy in pendulum 1, then back to potential energy in spring 2 and so on. Propagation of a compressional wave through a medium is, therefore, a process of converting the energy back and forth between kinetic and potential energy of the particles in the medium as the wave progresses.

When we compress a spring, or any elastic material, there is some internal friction, which causes some of the potential energy to be converted to heat. Also, when we set particles in motion, there will be some frictional resistance to the motion, which will convert some of the kinetic energy to heat. Because of these losses, which appear as heat in the medium, the wave will decrease in intensity as it propagates through the medium. We call this loss of intensity *attenuation*. The attenuation is a property of the medium and varies greatly from one medium to another, and in a given medium it often varies with the frequency of the wave.

When the wave reaches the last pendulum at the end of the line B in Fig. 1.12, it is obvious that the situation at this pendulum is different from that at the other pendulums. If there is no restraining spring beyond this pendulum it will be free to move to the right when pendulum 10 starts to compress spring 11. This will cause the spring to be stretched, which will, in turn, pull pendulum 10 so that a wave will progress back along the system toward A. We can say that the wave is *reflected* at the *termination* of the system and we can also say that the termination is open in this instance. Since the velocity of the wave is uniquely determined by the medium and the frequency does not change, the wavelength of the reflected wave will be the same as that of the forward-going wave. If the length of the system or the frequency of the source at A are adjusted so that the phase of the motions of the pendulums due to the reflected wave is the same as that due to the outgoing wave, the two motions will reinforce each other and the magnitude of the motions will build up until the rate of energy loss in the medium will be just equal to the rate at which energy is fed into the system at A. The wave pattern in the system under these conditions is called a *standing wave* pattern. Because the pendulum at B is unrestrained and a compression of spring 11 is reflected as a stretching or rarefaction of the spring we can say that the reflected wave is 180 degrees out of phase with the incident wave. Remember that in our illustration of simple harmonic motion in Fig. 1.2, one complete cycle corresponds to 360 degrees so, 180 degrees corresponds to one-half cycle difference in phase.

If we terminate the system by means of a solid wall against which the last pendulum is attached, the wave will again be reflected, but in this instance, because the last pendulum cannot move, the reflected wave will be reflected in phase with the incident wave, because when the last spring is compressed it will suffer a greater compression. A standing wave pattern can also be formed under these conditions, but because the reflected wave is now in phase with the incident wave it will be necessary to readjust either the length of the system or the frequency of the source to produce a standing wave pattern.

The vibration of the air column in the Kundt's tube is a standing wave with reflections from the closed end of the tube and the piston at the end of the rod. This is the reason for the necessity of carefully adjusting the position of the tube on its supports in order to observe the excitation of the powder in the tube. We will consider other applications of standing waves in later chapters.

The two types of termination described above represent two extremes. We could terminate the system with a different spring-pendulum system either by using different springs or different pendulums. If this is done there will be some reflection at the termination and some transmission through the termination. In order to avoid reflection at the termination, a quantity that we call the *acoustic impedance* of the medium must be the same on both sides of the termination. If the medium is identical on both sides the acoustic impedance will be identical. The acoustic impedance can be shown to be equal to the product of the density of the medium and the velocity of the wave in the medium. We could, therefore, terminate the pendulum-spring system of Fig. 1.12 with another pendulum-spring system with different mass in the pendulums and spring stiffnesses adjusted so that the acoustic impedances are matched. There will then be no reflection of the waves at the termination.

The pendulum-spring system shown in Fig. 1.12 is unique because the wave can be propagated only along the line of pendulums and springs. If there were no attenuation in the system, the intensity of the wave at the end of the system would be the same as that at the beginning. Sound waves propagated along a metal rod will behave in this manner.

The waves generated on the surface of water due to drops of water falling on the surface are different from sound waves in that the particle motion is perpendicular to the direction of propagation of the waves, but they illustrate the condition where the waves are confined to a plane. If we watch such an experiment the waves will appear as in Fig. 1.13 where O is the point where the waves

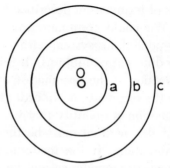

Fig. 1.13.   Wave crests propagating out from a disturbance on the surface of water.

are generated and a, b and c are successive crests of the wave traveling outward from the point O. It is obvious that as the wave propagates outward from O the energy of the wave will be distributed over successively greater lengths on the surface. The contours of the wave crests are arcs of circles so the length of surface over which the energy of the wave is distributed will depend on the distance r from O out to the contour. The length of the circumference of a circle is equal to $2\pi r$ so if r is doubled, the length of the circumference will be doubled and the amount of energy passing through a unit length per second, which we can define as the intensity, is halved. With waves confined to a plane, the intensity will be proportional to $1/r$, where r is the distance from the source. If there is attenuation in the medium, this will, of course, account for a still further reduction in the intensity.

Many sound sources are essentially point sources, which generate sound waves that propagate out from the source in all directions. The wave fronts from such a source will be spheres and the energy of the wave at some arbitrary distance from the source will be distributed over the surface of the sphere. The surface area of a sphere is equal to $4\pi r^2$, so the intensity, which can be defined as the energy passing through a unit area per second, will be proportional to $1/r^2$. If the distance r from the source is doubled, the intensity will be one-fourth as great, if there is no attenuation.

Standing waves can be established in a cavity. For example, standing waves can be set up in an air column in a tube. If the tube is open at one end and closed at the other, the longest wavelength that can be established as a standing wave in the tube is one which is four times the length of the tube. This will be treated in greater detail in the discussion of organ pipes in Chapter 11. Since a standing wave will build up until the rate of loss as heat is equal to the rate at which energy is supplied to the system, such an air cavity can be used to reinforce the sound from a source, and when used in such an application it is called a resonator. Helmholz used resonators to analyze sounds. The tone that can be heard when you listen to a seashell is at the frequency corresponding to the wavelength at which the cavity in the seashell resonates. In a total absence of external sound, there would be no sound from the seashell. It simply acts as a resonator to reinforce the frequency present in the external noise, which corresponds to the resonant frequency of the cavity in the seashell.

1.4. *Sound Intensity, Sound Pressure and Particle Velocity.*

We have already defined sound intensity as the amount of sound energy passing through a unit area per second. The unit of area will be determined by the system of units in use. If we are working in the metric system this unit of area would probably be a square centimeter (cm.). In the meter kilogram second (MKS) system, the unit of area would be one square meter and in the British foot pound second (fps) system it might be one square inch or one square foot. The unit of energy might be the erg or the Joule. If the Joule is used as the unit of energy, we might measure the intensity in Joules per square centimeter per second. Since the common unit for measuring power is the watt, and one watt is equal to one Joule per second, one Joule per square cm. per second is equal to one watt per square cm.

If we refer again to our simplified system in Fig. 1.12, we see that the individual pendulums move back and forth along the line AB as the wave progresses and the springs are alternately compressed and stretched. If we consider a medium such as air, the medium will be alternately compressed and rarefied and the particles (molecules) of the air will move back and forth parallel to the direction of propagation of the sound wave. When the medium is compressed its pressure will increase above the steady state pressure, and when it is rarefied its pressure will decrease to a value below that of the steady state pressure. The *sound pressure* is determined by the variation in pressure that takes place as the sound wave passes a point in the medium. The sound pressure is quite an important factor, since many of our sound-detecting devices—including our ears—respond to the sound pressure.

The motion of the particles of the medium can also be used as a means of detection of sound. One type of microphone, called the ribbon microphone, detects sound by responding to the particle velocity.

As the mass of the particles is increased, the magnitude of the motion and, therefore, the velocity of the particles will decrease. As the elastic stiffness of the medium is increased the magnitude of the sound pressure will increase. We have defined the acoustic impedance of the medium as the product of the density ϱ and the velocity of sound c in the medium. We have also indicated that Newton showed that the velocity of a wave in a medium is equal to the square root of the elasticity divided by the density. The acoustic impedance will therefore be equal to $\varrho\sqrt{E/\varrho}$, where the Greek letter ϱ is the symbol normally used to represent the density. The acoustic impedance will therefore be equal to $\sqrt{\varrho E}$.

The acoustic impedance is therefore an important factor relating sound intensity, sound pressure and particle velocity—i. e.,

$$I = p^2/\varrho c,$$

where I is the sound intensity, p is the sound pressure and ϱc is the acoustic impedance.

It is also possible to show that

$$I = pv,$$

where v is the particle velocity, so that

$$pv = p^2/\varrho c,$$

or
$$v=p/\varrho c.$$

If ϱc is low in value, v will be large relative to p and a particle velocity type of detector will be desirable. However, if ϱc is large, a pressure type of detector is more desirable. The acoustic impedance of air is 42 grams per square centimeter per second and that of water is 150,000 grams per square centimeter per second. Both pressure and velocity detectors are used successfully in air but it is very difficult to use a velocity detector in water.

At 0.0002 dynes per square centimeter, which is the threshold of hearing for the average human ear at 1000 cycles per second, the particle velocity will be
$$v=0.0002/42=4.75 \times 10^{-6}\text{cm. per second}$$
or 4.75 millionths of one cm. per second. The amplitude of motion, A, of the air particles is equal to $v/2\pi f$ or
$$A=v/2\pi f=4.75 \times 10^{-6}/2\pi \times 1000=4.75 \times 10^{-6}/6200$$
or
$$A=7.68 \times 10^{-10}=0.000000000768 \text{ centimeters.}$$
The diameter of a hydrogen molecule is 2.34×10^{-8} cm. so if we divide the diameter of a hydrogen molecule by this amplitude we will have
$$2.34 \times 10^{-8}/7.68 \times 10^{-10}=305.$$
The diameter of a hydrogen molecule is, therefore, approximately 300 times as great as the amplitude of particle motion at the threshold of human hearing at 1000 cycles per second. The ear is indeed a sensitive sound detecting instrument.

1.5. *Sound.*

Sound, in the strict sense, is a compressional wave that produces a sensation in the human ear. The average human ear will respond to waves in the frequency range from about 20 cycles per second to about 15,000 cycles per second.* The piano keyboard extends from the low a at 22½ cycles per second to the high c at 4186 cycles per second. The a above middle c has a frequency of 440 cycles per second. If we go up the frequency scale one octave, the frequency is doubled. For example , a_1 the lowest note on the piano, has a frequency of 22½ cycles per second and a_2, which is one octave above it, has a frequency of 55 cycles per second. The a_7 and a_8 are one octave apart at the upper end of the piano keyboard and their frequencies are, respectively, 1760 and 3520 cycles per second. The total frequency range of sensitivity of the average human ear is more than nine octaves. The science of acoustics is concerned with frequencies beyond the range of sensitivity of the human ear. We designate the frequency range below the audible range as the *infrasonic* range of frequencies, that range which is audible is called the *sonic* range and the range above the sonic range is called the *ultrasonic* range. Many animals utilize frequencies in the ultrasonic range.

---

* In this book I have expressed frequency in cycles per second, as I feel that it may be more meaningful to nontechnical readers. In the technical literature, frequency is now commonly expressed in Hertz (*Hz*) where 1 *Hz* equals 1 cycle per second.

2.

Units and Reference Quantities
Used in Measuring Sound

2.1. *The Decibel Scale for Measuring Sound Levels.*

We learned in the last chapter that sound is propagated through a medium as a compressional wave. The intensity of the sound is equal to the energy passing through a unit area per unit time and there is a pressure variation (the sound pressure) and a particle velocity variation (the particle velocity) that accompanies the propagation of the sound wave. It is difficult to measure the intensity of the sound directly, so the sound is generally measured by measuring either the sound pressure or the particle velocity. Because the sound intensity is related to the sound pressure by the relation, $I = p^2/\varrho c$, we can always compute the intensity if we know the sound pressure and the density ϱ of the medium and the velocity of sound in the medium c.

Microphones that respond to the particle velocity are often used in communication systems, but in sound measurements in both liquid and gas media pressure microphones are nearly always used. The various kinds of pressure and velocity microphones will be described in Chapter 6. The human ear responds to the sound pressure so we are often more concerned with the value of the sound pressure than we are with the intensity of the sound.

Most of the human senses, including our sense of sound, respond to the logarithm of the stimulus. If we listen to a given sound pressure, and the sound pressure is suddenly doubled, we have an indication by our ears of a certain *increase in loudness* that is the same regardless of the sound pressure of the original sound. Because of this, it is more convenient to use a scale based on sound pressure ratios, which we call a *decibel* (dB) scale (see Appendix 1 for a more rigorous derivation of the decibel scale). This scale is called a *sound pressure level scale* and there is then also a corresponding *sound intensity level scale* based on ratios of sound intensities. If you listen to the intensity of sound emitted by the flapping wings of a flea and then listen to that produced by two fleas, the intensity in the second instance will be double that in the first instance. If we listen to a cannon being fired and then listen to two cannons being fired simultaneously, the sound intensity again will be doubled. If we measured the intensities in these four instances, the intensity difference between the sound of one cannon and two cannons would be vastly greater than the intensity difference between the sound of one flea and two fleas. The increase in loudness in the two instances will, however, be the same. We can arrange our dB scale to handle this problem if we define it so that a sound intensity 10 times as great as another sound intensity differs from it by 10 dB and then divide the intervening scale so that, regardless of the position in the scale, doubling the intensity ratio always produces the same dB difference. Figure 2.1 shows an example of such a scale. This scale

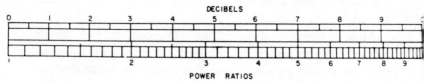

Fig. 2.1.   A decibel scale for measuring intensity ratios.

has the further advantage that it can be used over any range of intensity ratios. If the intensity ratio falls between 10 and 100 units, the same scale shown in Fig. 2.1 will extend from 10 to 20 dB.

Since we more commonly measure sound pressures, it is desirable to express ratios of sound pressure levels on a dB scale, and furthermore this scale should be so arranged that, if we read the number of dB representing a given sound pressure ratio, it will be equal to the number of dB representing the corresponding sound intensity ratio. The sound intensity is proportional to the sound pressure squared so we arrange the dB scale for sound pressure ratios so that 20 dB corresponds to a factor of 10 in sound pressure ratio. The number of dB representing a given sound pressure ratio will then be equal to the number of dB representing the corresponding sound intensity ratio. The analysis in Appendix 1 shows why this is true. Figure

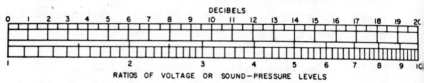

Fig. 2.2.   A decibel scale for measuring pressure or voltage ratios.

2.2 shows the scale for converting sound pressure ratios to dB. Figures 2.1 and 2.2 can be used directly for converting sound intensity ratios and sound pressure ratios to dB. If a ratio of sound pressure, for example, is less than one, say 1/6, we can simply invert the fraction which will be 6/1=6 and read the number on the sound pressure dB scale and record it with a negative sign. In the above example, we would read the dB value corresponding to 6 on the sound pressure level scale, which is 15.5 dB. The level of the ratio 1/6 is therefore equal to −15.5 dB.

In order to extend the use of the scales to any conceivable sound pressure ratio or sound intensity ratio, Table 2.1 shows the readings on the two ends of the dB scales corresponding to a wide range of pressure and intensity ratios.

Since the decibel scale is based on ratios, it is customary to establish certain reference quantities and to express on the dB scale the ratio of the measured quantity to the reference quantity. In air acoustics, the reference sound pressure is chosen as the lowest sound pressure that can excite the hearing sense in an average human ear. This corresponds to 0.0002 (2×10^{-4}) dynes per square cm. sound pressure or 0.0000000000000001 (10^{-16}) watts per square centimeter sound intensity. We can then express some measured sound pressure in dB if we take the ratio of that sound pressure to the reference sound pressure and convert that ratio to dB by means of Fig. 2.2 and Table 2.1 to obtain the value of the sound pressure level in dB relative to the reference value. Figure 2.3 shows the sound level in dB, referred to the threshold of hearing, of a number of kinds of sounds with the corresponding sound pressures and sound intensities.

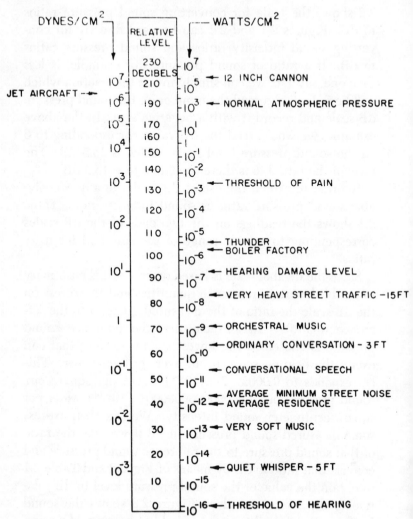

Fig. 2.3. Chart showing values of sound intensity and sound pressure of various kinds of acoustic signals with the corresponding decibel values relative to 0.0002 dynes per square centimeter.

2.2.   *Use of dB Scales in Measurements.*

In order to measure sound levels, it is necessary to convert the sound pressure to an electrical signal that can be amplified and measured. A microphone is a device that performs the function of converting the sound energy to electrical energy. When a microphone is used for sound

Table 2.1
DECIBEL RANGE FOR VARIOUS SOUND INTENSITY AND PRESSURE RANGES

dB range	Sound Pressure range	dB range	Sound Intensity range
0–20	$1-10$	0–10	$1-10$
20–40	$10-100$	10–20	$10-100$
40–60	10^2-10^3	20–30	10^2-10^3
60–80	10^3-10^4	30–40	10^3-10^4
80–100	10^4-10^5	40–50	10^4-10^5
100–120	10^5-10^6	50–60	10^5-10^6
120–140	10^6-10^7	60–70	10^6-10^7
140–160	10^7-10^8	70–80	10^7-10^8
		80–90	10^8-10^9
		90–100	10^9-10^{10}
		100–110	$10^{10}-10^{11}$
		110–120	$10^{11}-10^{12}$
		120–130	$10^{12}-10^{13}$
		130–140	$10^{13}-10^{14}$
		140–150	$10^{14}-10^{15}$
		150–160	$10^{15}-10^{16}$

measurements, it is calibrated to determine its *sensitivity*. The sensitivity of a microphone is defined in terms of voltage generated per unit of sound pressure. The voltage is expressed on a decibel scale relative to one volt as the reference voltage and, since the electrical power is proportional to the voltage squared, the voltage can be converted to decibels relative to one volt by means of Fig. 2.2 and Table 2.1. The sensitivity of a microphone is usually given in terms of dB relative to one volt per dyne per square cm. (dBV per dyne per square cm.). Here we have a slight problem. The acousticians usually insist on expressing the sound pressure in dB relative to 0.0002 dynes per square cm. but they calibrate their microphones in terms of one dyne per square cm. which is 74 dB above the threshold of hearing. (You can verify this with the dB scale in Fig. 2.2 and Table 2.1 if you find the number of dB corresponding to 1/0.0002.)

We are now about to discover another advantage of the decibel system. Let us assume that we have a signal which is amplified by two amplifiers in series, as indicated in Fig. 2.4. The signal voltage would be multiplied

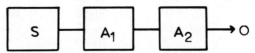

Fig. 2.4. Schematic of an amplifier system for amplifying the electrical signals generated in a microphone.

by the gain in the amplifier A_1 and the value obtained would then be multiplied by the gain in amplifier A_2 to obtain the value of the voltage output O. We can, however, express the voltage gain in an amplifier in dB by taking the ratio of the output to the input and converting this to dB by use of Fig. 2.2 and Table 2.1. Since the dB scale is a logarithmic scale, a product is obtained by adding the dB values.

If the sensitivity of the hydrophone is —80 dB relative to one volt per dyne per square cm. (—80 dBV per dyne per square cm.), the gain in amplifier A_1 is 40 dB, the gain in amplifier A_2 is 100 dB and the acoustic sound pressure is 34 dB relative to 0.0002 dynes per square cm., the output at O will be

signal level —sensitivity —conversion to 0.0002 +gain in A_1 +gain in A_2.

or

$$34 - 80 - 74 + 40 + 100 = -154 + 184 = 20 \text{ dBV.}$$

Normally we use such a system to determine the value of an unknown signal from the reading of the output voltage of the system. In this instance the signal level will be

Signal level = Output (dBV) + sensitivity + conversion to 0.0002 — gain in amplifier A_1 — gain in amplifier A_2

or

$$34 \text{ dB} = 20 + 80 + 74 - 40 - 100.$$

This would correspond to the sound pressure level of ordinary conversation. (see Fig. 2.3).

We have not considered the effect of frequency of the sound or the bandwidth of the signal. When a microphone is calibrated at a calibration facility, the sensitivity as a function of frequency is normally plotted. Most microphones used for sound measurements have quite uniform sensitivity as a function of frequency over the sonic range. However, we are often interested in measuring the signal levels in limited regions of the spectrum. In order to do this, filters can be incorporated in the amplifiers used to amplify the signal received by the microphone. The sound power will be proportional to the band width measured in cycles per second if the distribution of sound energy is constant over the entire band width. The scale in Fig. 2.1 will be applicable to determining the intensity in one band width from a measurement in another band width. For example, if the level of the signal is measured in a 100

cycle per second band width, the level over a 10 cycle per
second band width will be 10 dB less and the level in a
one cycle per second band width, usually designated dBs,
will be less by another 10 dB. It is sometimes desirable,
when measuring noise levels, to measure the noise over
the entire sonic region and also to utilize filters in the
amplifiers to measure the noise in certain restricted bands.
Reference to Fig. 2.3 will indicate the tremendous range
of intensity over which the human ear can function.

The bar, which is equal to one million dynes per square
centimeter, is equal to 0.98697 of a standard atmosphere.
One dyne per square centimeter is, therefore, approximately equal to one millionth of a bar (one microbar)
which is usually written one μbar, where the Greek letter
μ is the abbreviation for micro.

Appendix To Chapter 2

THE DECIBEL SCALE

In Chapter 2, the method of use of a decibel scale was
developed without reference to the true mathematical
basis on which it is founded.

The decibel scale is based on the properties of the log-
arithms of numbers. A logarithmic scale has three important advantages. It is convenient for dealing with
large changes in the variables, it greatly simplifies many
of the computational processes and the response of the
human organism to many stimuli, including sound, is approximately logarithmic.

The fundamental division of a logarithmic scale for expressing the ratio of two amounts of power is the Bel. The
logarithm, to the base 10, of the ratio of two amounts of
power is defined as the number of Bels.

If we are considering an amount of power, P, and P_o is a reference power, the number of Bels denoting the ratio P/P_o is

$$N = \log_{10} (P/P_o) \text{ Bels.} \qquad (1)$$

If P is 10 times P_o, the ratio P/P_o will be one Bel. This unit is too large for convenience, so the decibel (dB) equal to 1/10 of a Bel is commonly used. Equation (1) then becomes

$$n = 10 \log_{10} (P/P_o), \qquad (2)$$

where n is the power ratio expressed in dB.

The sound pressures at some point resulting from the two amounts of power, P and P_o, can be expressed as

$$P = p^2 A/\varrho c \qquad (3)$$

and

$$P_o = p_o^2 A/\varrho c \qquad (4)$$

so

$$P/P_o = (p/p_o)^2. \qquad (5)$$

The logarithm of the square of a quantity is equal to two times the logarithm of the quantity so,

$$10 \log_{10} (P/P_o) = 20 \log_{10} (p/p_o) \text{ dB.} \qquad (6)$$

Therefore, when we wish to express the ratio of two sound pressures on a dB scale, we must take 20 times the logarithm of the ratio of the sound pressures in order to arrive at the same dB value for the sound signal as we would if we had considered the sound power or the sound intensity. This is the reason for the difference in the scales shown in Figs. 2.1 and 2.2. Since sound intensity $I = P/A$, we can write

$$P/P_o = I/I_o, \qquad (7)$$

so

$$10 \log_{10} P/P_o = 10 \log_{10} I/I_o. \qquad (8)$$

Since the decibel scale is designed to express ratios of quantities such as sound pressure, sound power or sound intensity, it is convenient to establish reference quantities to which absolute values can be referred. The decibel scale is also used to represent the ratios of such quantities as electric power and voltage. For example: a voltage amplifier develops an output voltage as a result of an input voltage, so we can express the voltage gain of the amplifier in dB as

$$dB = 20 \log_{10} (E_{out}/E_{in}). \qquad (9)$$

The decibel value for a pressure level p relative to a reference level, p_0, will be

$$dB \text{ relative to } p_0 = 20 \log_{10} (p/p_0). \qquad (10)$$

The value of the sound intensity at the threshold of hearing for the average human ear is approximately 10^{-16} watt per cm^2 which corresponds to a sound pressure of 0.0002 dyne per cm^2.

A sound pressure of one dyne per cm^2 relative to 0.0002 dyne per cm^2 can be expressed in dB as

$$dB (p_0) = 20 \log_{10} (1/0.0002)$$

or

$$dB (p_0) = 20 (0.000 - \overline{4}.3096)$$

or

$$dB (p_0) = (20) (3.6904) = 73.8.$$

A sound pressure of one dyne per cm^2 is therefore 73.8 dB relative to 0.0002 dyne per cm^2. The value 74 dB is usually taken as a sufficiently close approximation. This is an important constant to remember, as we often must convert the dB value of sound pressure ratios from one of these reference values to the other. Most work in air acoustics involves the use of 0.0002 dyne per cm^2 for measurement but calibrations are often done to the reference pressure of one dyne per cm^2. In most work in underwater acoustics, however, the reference sound pressure is one dyne per cm² for both measurement and calibration.

When we express sound pressures, intensities or powers on a decibel scale, we call them sound pressure *level*, sound intensity *level* and sound power *level*. It is important to always state the reference value to which any sound pressure level, sound intensity level or sound power level is referred.

3.

Speech and Hearing

3.1. *Speech.*

We can think of the speech mechanism as consisting of a power supply, a vibrating element and a system of valves and resonators.

The breathing mechanism supplies the power for driving the speech mechanism. It consists of the lungs and the rib cage with its associated muscles and it operates as an air pump to drive air through the vibrating element to supply the power necessary to drive it. The power supplied to the vibrating element by the breathing mechanism is equal to the product of the difference in pressure on the two sides of the vibrating element and the volume of air which passes through it.

The vibrating element, called the *vocal chords* or *vocal folds,* is located in the *larnyx.* The vocal folds are long, smooth rounded bands of muscle tissue which may be lengthened, shortened, tensed or relaxed. They can be separated or drawn together. During normal breathing the folds are widely separated.

Vocal sounds are produced by drawing the folds together, causing the air pressure to build up below them so that it blows the folds apart until the pressure below the folds is reduced sufficiently so that they will again be drawn together. Each time the folds are blown apart, a pulse of pressure is generated and the frequency of the pulses is determined by the tension in the folds. If the vocal folds are abruptly contracted the vibration will be stopped. This is called a *glottal stop*. If the contraction is abruptly released the vibration will suddenly start again. This is called a *glottal attack*. A normal attack is achieved by releasing the contraction more gradually.

The pitch of the sound produced is controlled by controlling the tension in the folds and the sound power is controlled by controlling the power supplied by the breathing mechanism. These various controls are used in the production of the sound for speaking and singing. An important part of the training of a singer is the refining of the controls over the vocal folds and the sound power.

The sound produced by the vocal folds is not a pure tone but instead consists of a fundamental and several harmonics. The center of the band of frequencies produced by the average female is about 260 cycles per second (approximately middle c) and that produced by the average male is about an octave lower.

The character of the sound that emerges from the mouth is greatly influenced by the various cavities in the head, including the cavities formed in the mouth. Without the influence of these resonators the sound produced would be quite unmusical. The speaker, or singer, can control the volume of the mouth cavity, and therefore has control over its resonant frequency. The frequencies of resonance of the various cavities determine which frequencies in the sound produced by the vocal folds will be emphasized. The quality of the voice sounds is therefore considerably modified by the resonators in the head. Everyone has noticed the change in voice quality in a person who has a head cold. Under these conditions, the tissues in the sinus cavities and in the nose are swollen so the resonant frequencies of these cavities are changed.

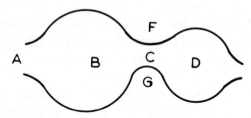

Fig. 3.1.   Formation of the cavities in the mouth with the tongue to produce vowel sounds. *A* is the mouth opening, *B* is the cavity in the front portion, *D* is is the cavity in the back portion, *C* is the channel coupling the cavities, *G* is the tongue and *F* is the roof of the mouth.

The two cavities which are of primary importance in producing the vowel sounds are formed in the mouth and are indicated schematically in Fig. 3.1. *A* is the mouth opening, *B* is the cavity in the front portion of the mouth and *D* is the cavity in the back of the mouth. These cavities are joined by the channel *C* formed by the tongue *G* and the roof of the mouth *F*. The speaker, or singer, controls the relative sizes of these cavities and the coupling between them by moving the tongue and lips.

3.2. *Hearing.*

The human ears serve to detect sounds, analyze them, and determine the direction from which they come. The physicist is generally concerned with the detection and measurement of sounds by physical means. However, the response of human ears to sounds determines whether they are considered pleasant or objectionable. The physicist is interested in determining the physical characteristics of sounds and coordinating these observations with the observations of trained listeners. There are trained listeners in many fields. An orchestra conductor can listen to the combined sounds emitted by the instruments in a large orchestra and he can tell very quickly if one of the performers plays off pitch or does not sound a note at the precise time that it should be sounded. A skilled automobile mechanic can listen to the jumble of sound from an automobile engine and diagnose its defects. It would be difficult, if not impossible, to construct physical measuring systems that could replace trained listeners in the many instances where they can function so effectively. This is partly due to the fact that the ear with the associated analysis system in the brain is such a marvelous system and it is also partly due to the fact that it is difficult for the trained listener to convey to the physicist who is not a trained listener exactly what he listens for. For this reason a physicist who wishes to learn about the physics of musical instruments, for example, should also be a musician. Professor Saunders contributed a great deal to our knowledge of violins because he was a physicist and also an accomplished violinist.

The human ear consists of three principal parts, the outer ear, the middle ear and the inner ear. These are shown schematically in Fig. 3.2. The outer ear consists of the *concha* or shell and the *meatus* or the tube leading from the outside to the ear drum, which separates the outer ear from the middle ear. Sound energy entering the ear travels through an air medium until it reaches the ear drum, which vibrates in response to the sound pressure.

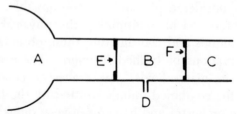

Fig. 3.2.   Schematic diagram of the ear. *A* is the outer ear, *B* is the middle ear, *C* is the inner ear or cochlea, *D* is the eustachian tube, *E* is the ear drum and *F* is the oval window.

The middle ear contains three tiny bones commonly called the hammer, anvil and stirrup which couple the vibrations of the ear drum to the inner ear.

The inner ear, called the *cochlea,* contains the sensing elements and is filled with liquid. It is necessary, therefore, for the middle ear mechanism to transmit sound from an air medium to the liquid medium in the inner ear. We saw in the previous chapters that the acoustic impedance of a liquid is higher than that of air, and therefore, for a given sound intensity, the sound pressure in the liquid medium will be higher than that in the air medium. The particle velocity will be correspondingly less in the liquid medium than in the air medium.

The bones in the middle ear are designed to make the necessary impedance transformation by behaving as a lever system to increase the magnitude of the force and decrease the magnitude of the motion. Figure 3.3 shows schematically how this is accomplished. The ratio of pressure exerted by the stirrup to that exerted on the hammer by the ear drum is about 25 dB.

The inner ear or cochlea is in the form of a tube coiled

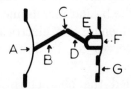

Fig. 3.3. Schematic diagram of the middle ear. A is the ear drum, B is the hammer, C is the hinge between the hammer and anvil, D is the anvil, E is the stirrup, F is the oval window and G is the round window.

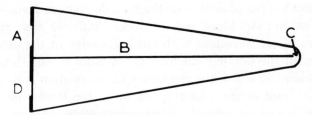

Fig. 3.4. Schematic diagram of the inner ear as it would appear if uncoiled. A is the oval window, B is the basilar membrane, C is the helicotrema and D is the round window.

in a manner similar to a snail shell. Figure 3.4 shows the cochlea diagrammatically as it would appear if it were uncoiled. There are two windows at the large end of the cochlea. The window to which the stirrup is attached and which is, therefore, the point at which the sound energy enters the cochlea is called the *oval window*. The other window at the large end of the cochlea is called the round window. The cochlea is divided by a membrane called the *basilar membrane* and the sound waves travel from the oval window along the basilar membrane to the small end of the cochlea, where there is a gap between the end of the basilar membrane and the end of the cochlea called the *helicotrema,* and then back to the round window on the other side of the basilar membrane. The elements that convert the sound energy to electrical energy for transmission by the nerves to the brain are located in the basilar membrane.

Helmholz studied the characteristics of the ear and concluded that the elements along the basilar membrane are tuned to vibrate at different frequencies, and therefore the particular elements that respond to a given frequency are the ones tuned to that frequency. This theory will account for our ability to distinguish the various frequencies of sound. More recent studies indicate that this theory is not consistent with our knowledge of the physiology of the hearing mechanism. A more recent theory by Békésy, in which the frequency discrimination is based on the form of the vibration set up in the basilar membrane, is now considered a better explanation. Our understanding of the functioning of the complete hearing system including the brain centers involved is far from complete.

Our hearing system is extremely sensitive and versatile. The lowest sound power that can be detected by an average ear is somewhat less than the lowest level of light power that can be detected by our eyes. The range of frequencies to which the average ear is sensitive is about nine octaves. It is extremely difficult to design electronic communications equipment capable of operating over such a range. The eye, which is sensitive over a frequency band of about two octaves, is not capable of recognizing the individual frequencies, but rather sees a color that may be a single optical frequency or a combination of two or more. The ear hears the individual frequencies, and because of this ability we are able to distinguish among the various frequencies of sounds that may be heard by utilizing this frequency discrimination.

Because we have two ears we are able to determine the direction from which a sound is coming and we utilize the directional characteristics of our two ears to assist us in discriminating a weak sound in the presence of a higher level sound field.

Our ears have their maximum sensitivity to sound at about 2000 cycles per second and they are about 70 dB less sensitive at 20 cycles per second and about 20 dB less sensitive at 10,000 cycles per second. The high frequency sensitivity decreases with age so that people over 60 years of age generally have a considerably reduced sensitivity above 5000 cycles per second.

When we measure the loudness level of a tone the reference intensity may be taken as the intensity of the hearing threshold of that frequency. If a tone of a certain frequency has an intensity I, its loudness can be given by the number of dB relative to the threshold intensity of that frequency. However, in order to make the reference intensity quite definite, the value for the threshold for an average ear at 1000 cycles per second has been chosen as the reference level. This intensity corresponds to a sound pressure of 0.0002 μ bar. This unit of loudness is called the *phon* and one phon represents approximately the least change in loudness that can be detected by the human ear under average conditions.

The ear is more sensitive in perceiving changes in pitch than changes in intensity. Under favorable conditions a 5% change in intensity is necessary to be perceived, while a 0.2% change in frequency can be perceived by a trained ear. The ratio of the minimum detectable change in frequency to the frequency is approximately constant over the range from 500 cycles per second to 4000 cycles per second.

The phon scale has the disadvantage of not giving a reliable impression of relative loudness. The *sone* scale was therefore designed to overcome this disadvantage. The loudness level of 40 phons is defined as having a loudness of one sone and a sound of n sones loudness is n times as loud as one of 1 sone. An increase of 10 phons is approximately equal to doubling the number of sones. Table 3.1 shows the relation between phons and sones.

Exposure to high sound levels over extended periods can damage our ears. Care should be exercised when sound levels approach 85 dB above a μ bar and higher. The tendency to hearing damage depends on the level of the sound and the length of exposure to it. When people are exposed to continuous sound fields in the 90 to 100 dB range, the possibility of hearing damage can be greatly minimized if they will take one or two breaks during each half day or, better still, wear suitably fitted ear protectors to reduce the sound level at the ear drum.

Legislation has been enacted in some states for the protection of hearing of workers in noisy industry. In order to properly enforce such legislation, it is necessary to measure the sound levels to which the workers are exposed and to measure the hearing sensitivity, over the audio range, of the workers at the initiation of their employment and periodically during their employment. A curve showing the sensitivity of the ears as a function of frequency is called an audiogram.

The problem of properly making audiograms for a large number of individual employees is a formidable one because of the skill required to make the measurements and

Table 3.1
Relation between phons and sones.

Phons	30	40	50	60	70	80	90
Sones	0.5	1	2	4	8	16	32

to keep the necessary instruments in proper repair and calibration.

Rock 'n' roll groups, which use instruments that incorporate electronic amplifiers, are a new threat to the hearing of young people. These orchestras often generate sound levels higher than 120 dB above a μbar. Even relatively short exposure to such sound levels—if it occurs quite often—can cause permanent hearing damage.

People who do considerable target shooting should wear ear protectors, as the high sound pressures generated by a gun being fired can cause permanent damage to the ears.

4.

Sound Propagation

4.1. *Divergence.*

We saw in Chapter 2 that if a pulsating source is radiating sound power, P, the intensity at a spherical surface of radius r from the source will be $I = P/4\pi r^2$. If the radius of the sphere is unity, the intensity I_0 will be equal to

$$I_0 = P/4\pi. \tag{1}$$

If we go to a radius r_1, the intensity at the surface will be

$$I_1 = P/4\pi r_1^2. \tag{2}$$

The ratio of the intensity at the radius r_1 to that at unit radius will be

$$I_1/I_0 = 1/r_1^2. \tag{3}$$

Since the ratio of intensities varies inversely as the square of the range, we can apply the decibel scale in Fig. 2.2 and Table 2.1 to determine the dB ratio of intensities corresponding to any ratio of ranges from the source. The reference radius or range is usually taken as one yard or one meter. To find the dB ratio of intensities for ranges greater than one yard, we use the scale and table to determine the dB value of $r_1/1$ or r_1. The decrease in intensity going from r_0 to r_1 will be minus the number of dB thus obtained. At 100 yards, for example, the change in intensity due to divergence will be $- 40$ dB. If the intensity I_0 is 10 dB relative to one watt per square cm, at one yard, the intensity $I_1 = 10$ dB $- 40$ dB $= - 30$ dB relative to one watt per square cm. or 0.001 watt per square cm. It is obvious that each time the range is doubled the intensity will decrease, due to divergence, by 6 dB. When we go from one yard to two yards, the intensity will decrease by 6 dB and when we go from one mile to two miles it will also decrease by 6 dB.

4.2. Reflection.

If a sound wave front encounters a surface where there is a great change in the value of the acoustic impedance, it will be reflected. If the surface is at a reasonably large distance from the source of the sound, the radius of the spherical wave fronts will be large and the curvature of the wave fronts will be so slight that very little error is instroduced if we assume that they are planes.

We can define a *ray* as any line drawn perpendicular to a wave front, as indicated in Fig. 4.1. As the distance from the source is increased the wave fronts approach more nearly to plane surfaces and the rays become more nearly parallel to each other.

Figure 4.2 shows how a wave front is reflected from a plane reflecting surface. The rays a, b and c are the incident rays and a', b' and c' are the refllected rays. The angle the reflected rays make with the perpendicular to the reflecting surface R is equal to the angle the incident rays make with the perpendicular to the surface. This is a universal law of reflection of waves from a surface. You can verify this law for light by means of a plane mirror.

If the reflecting surface is curved, the law of reflection for each individual ray will still be the same, but since the

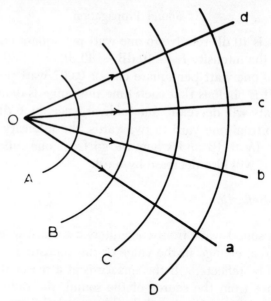

Fig. 4.1. Wave fronts and rays from a source of sound O. The surfaces A, B, C and D are the wave fronts and the lines a, b, c and d are the rays.

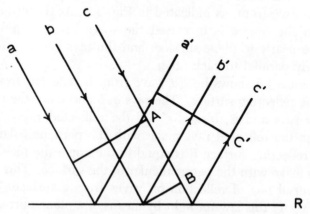

Fig. 4.2. Reflection of a plane wave from a plane surface, R. The rays are a, b and c and A and B are wave fronts before reflection; a', b', and c' are the rays and C' is the wave front after reflection.

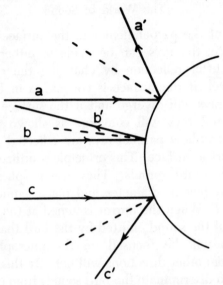

Fig. 4.3. Reflection of the rays of a plane wave front from a convex surface; *a*, *b* and *c* are the incident rays and *a′*, *b′* and *c′* are the reflected rays. The dotted lines are the perpendiculars to the surface.

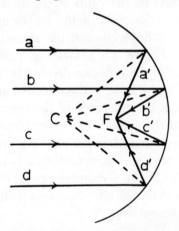

Fig. 4.4. Reflection of parallel rays from a concave spherical reflector; *a*, *b*, *c* and *d* are the incident rays and *a′*, *b′* *c′* and *d′* are the reflected rays. The dotted lines are the perpendiculars to the surface.

direction of the perpendicular to the surface varies over the surface, the rays can be made to either diverge or converge after reflection by changing the curvature of the surface. If the surface is convex, as in Fig. 4.3, the reflected rays will diverge and if the surface is concave, the reflected rays will converge as shown in Fig. 4.4, which shows how parallel rays are reflected from a concave spherical surface. This principle is utilized by people who record bird sounds. They use a spherical mirror about four feet in diameter and place a microphone at the point F. When the mirror is aimed at the bird, essentially all of the sound emitted by the bird that the mirror intercepts will be focused on the microphone, while sounds from other directions will not. By this means, it is possible to discriminate the bird sounds from other sounds always present in the woods.

If the reflector is small compared to a wavelength its effective area as a reflector becomes less than its actual area. The value of the ratio of the effective area to the actual area as a function of the wavelength was derived theoretically by Rayleigh and has since been verified experimentally. Figure 4.5 shows the value of $\pi d/\lambda$ (the circumference of the reflector divided by the wavelength) plotted against the ratio of the effective area to the actual area. Here d is the diameter of the reflector and λ is the wavelength of the sound. If the wavelength of the sound is equal to the circumference of the reflector the ratio of the effective area to the actual area is essentially unity, but if the ratio of the circumference to the wavelength is decreased to 0.1, the effective area will be only about 0.001 of the actual area.

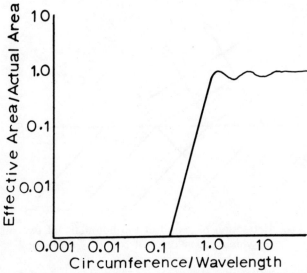

Fig. 4.5.   Curve showing how the effective area of a small
reflector depends on the ratio of the circumference
of the reflector to the wavelength of the sound.

4.3.   Refraction.

When a wave front of any kind of wave passes from a
medium where the velocity of propagation is c_1 to another
medium where the velocity of propagation is c_2, the direction of the rays will be changed unless they are perpendicular to the surface. You can observe this effect with
light very readily by holding a straight stick so that part
of it projects under water at an angle to the surface. The
stick will appear to be bent at the interface of the air and

the water. This is caused by the fact that the velocity of light is less in water than in air.

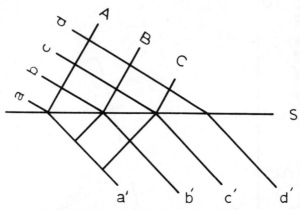

Fig. 4.6. Refraction of sound at the boundary of two media in which the velocity of sound is different.

Figure 4.6 shows the paths of rays across the interface of two media in which the velocities are c_1 and c_2. The velocity of the waves, c_2, in the lower medium is less than the velocity, c_1, in the upper medium. As the wave front A crosses the boundary into the lower medium it will be slowed, causing the rays a, b, c and d to be bent as the wave front crosses the boundary S. If the wave originates in the medium of velocity c_1, with the ray directions a, b, c and d, the ray directions in the medium of velocity c_2 will be a', b', c' and d' and if the wave originates in the medium of velocity c_2 with ray directions a' b' c' and d' the ray directions in the medium of velocity c_1 will be a, b, c and d. The law of refraction at the interface of two media was developed by Snell. However, to consider Snell's law quantitatively it is necessary to utilize trigonometry. This is done in the Appendix to this chapter.

We may have two homogeneous media separated by a sharp boundary but we are more often concerned with propagation of sound over great distances in air or water where the velocity varies continuously in the vertical direction in the medium. Snell's law still holds in this instance. For example, referring to Fig. 4.6, if the sound originates in a medium where the velocity is c_1 with ray directions as indicated by a, b, c and d, the direction of the rays in the medium of velocity c_2 will be uniquely determined by the velocities c_1 and c_2 and the directions of the rays a, b, c and d regardless of whether the velocity changes abruptly at the surface S or changes gradually over a large distance. If the velocity changes gradually over a large distance, the rays will follow curved paths over this distance.

The velocity of sound in a gas varies as the square root of the absolute temperature, so as the temperature decreases the velocity of sound decreases. We saw from Fig. 4.6 that the rays are bent toward the region of lower velocity, so if the temperature decreases as we go up in the atmosphere, sound rays generated near the surface of the earth will be refracted upward and this refraction will play an important role in sound propagation in the air. Under some weather conditions there will be a temperature inversion at some altitude above the surface of the earth; that is, there will be some altitude at which the temperature increases. Such a temperature inversion can be great enough so that the rays will again be bent downward so that a sound such as a cannon shot or an explosion may be heard at some great range but can not be heard at some intermediate range.

The effect of refraction on sound propagation in the

ocean is very important and will be considered in detail
in Chapter 13.

4.4. *Attenuation.*

We have mentioned before that when energy is propagated as waves through a medium, some of the energy is
converted to heat in the medium and is therefore lost.
This loss of energy is in addition to the decrease in intensity due to divergence. When sound is propagated through
the air, it is attenuated. The attenuation varies as a function of the frequency and also as a function of the relative
humidity of the air. Figure 4.7 shows how the relative
humidity at which the absorption is a maximum varies
with frequency.

We can express the attenuation in dB per yard or dB

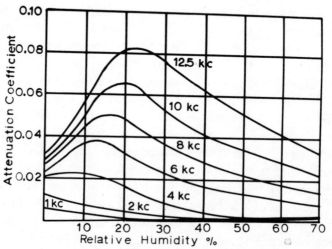

Fig. 4.7.   Attenuation coefficient of air for sound at different
frequencies and relative humidity.

per meter. If there is a reduction in intensity in one meter of 0.04, this can be expressed in dB by taking the ratio of the intensities at the beginning and end of a meter, which is $1.00/0.96 = 1.042 = 0.18$ dB. Since the fractional reduction in intensity is the same in each successive meter, the attenuation in r meters would be $r \times 0.18$ dB. The Greek letter α is usually used to designate the attenuation in a unit distance, so the attenuation over any range is r α dB.

We can express the intensity I_r at any range due to an initial intensity I_o as $I_r = I_o$ — dB reduction due to divergence — $r \alpha$. However, this calculation does not take into consideration the loss in intensity due to refraction. The effect of refraction is, of course, negligible over short ranges but over large ranges the refraction of sound may be the principal factor determining the level of sound received.

There are not many applications of long-range sound propagation in air, but there is considerable utilization of this principle under water. We will consider this in detail in Chapter 13, where we will also consider the propagation of sound in the sediment layers under the sea and show how we can determine the structure of the layers beneath the sea by means of sound propagation measurements.

Appendix to Chapter 4.

SNELL'S LAW

If the velocity of sound is known at all points in a medium containing a sound source, we can trace the path of any sound ray in the medium by the application of Snell's law which gives the relation between the angle of the ray and its velocity. Snell's law can be expressed mathematically as

$$c_1 \cos \theta_1 = c_2 \cos \theta_2 = c_3 \cos \theta_3 = c_n \cos \theta_n.$$

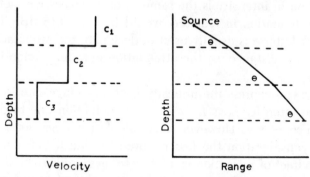

Fig. A1. Application of Snell's law to the case of three isothermal layers.

Figure A1 shows a layered medium where c_1, c_2 and c_3 are the sound velocities in the three layers and θ_1, θ_2 and θ_3 are the angles which the ray makes with the horizontal in each of the layers. The left-hand portion of Fig. A1 shows the velocity as a function of depth in the medium and the right-hand portion of the figure shows the path of a ray in the medium. In Fig. A1, it is assumed that the three layers are individually isovelocity. However, the direction of a ray at a point is determined uniquely by its direction at the source, the velocity of sound at the source and the velocity at the point in question regardless of the way in which the velocity varies between the source and the point in question.

If, at some region in the medium, the velocity reaches a value c_x such that

$$c_x = c_0 / \cos \theta_0,$$

where c_0 is the velocity at the source and θ_0 is the angle
that the ray makes with the horizontal at the source, it
indicates that at that point the direction of the ray is horizontal since $\cos \theta = 1$. The ray will not be able to penetrate in depth beyond that point but will be refracted in
depth back toward the depth from which it originated.

5.

Noise

To a gardener, a weed is any plant growing out of place.
Using this analogy, we can define noise as any sound undesirable at a particular time and place. A Beethoven
sonata when played in the presence of a person trying to
concentrate on some problem is noise to him.

In a more technical sense, we can define noise as an
erratic, intermittent or statistically random oscillation.
The sounds emitted by motor vehicles and machinery are
examples of such noise. It is often possible to represent
a noise as an infinite sum of sinusoidal vibrations, the intensities, frequencies and phase relations of which are
such that they add up to the given noise.

Random noise is an oscillation whose magnitude is not
specified for any given instant. The instantaneous magnitudes of a random noise are specified by the probability
of the fraction of the total time that the magnitude lies
within a specified range. A random noise whose instantaneous magnitudes occur according to the Gaussian distribution is called a "Gaussian random noise." A Gaussian
distribution is a theoretical intensity distribution, used in
statistics, that is bell-shaped, symmetrical and of infinite
extent.

We use the term "white noise" to describe a noise
having a uniform distribution of energy as a function of

frequency over the sonic range. We sometimes use the term to describe noise in other regions of the spectrum or over limited regions. For example, we can speak of a white noise over the region from 1000 cycles per second to 5000 cycles per second.

Depending on the physical nature of the source, a noise field may be random or white; it may contain specific frequencies or it may have a relatively narrow band of frequencies. The noise in a ventilating system has a background of white noise due to the flow of air in the pipes, with some specific frequencies superimposed such as those emitted by the motor driving the fan, the frequency of rotation of the fan and the product of the frequency of rotation of the fan and the number of blades on the fan.

Ambient noise is the noise that exists in the medium because of uncontrolled sources such as the traffic in a street or the wind blowing through trees in the woods.

We sometimes speak of self noise and external noise. In the application to a motor vehicle, for example, the self noise is the noise heard by an observer riding in the vehicle while the external noise is the noise heard by an observer outside the vehicle as it passes.

When we deal with noise, it is necessary to analyze it to determine its distribution with frequency and to determine if discrete frequencies are present. In some instances we are interested only in the total level of sound, while in others we are interested in the frequency distribution of the levels of the noise. We saw in Chapter 2 that the intensity level in dB for a white noise signal of band width $(\Delta f)_1$ relative to another signal of band width $(\Delta f)_2$ can be found by determining the dB value of the ratio of $(\Delta f)_1/(\Delta f)_2$ using the scale in Fig. 2.1 and Table 2.1.

When we deal with noise in practice we may be interested only in the intensity level of the total spectrum of the noise or we may be interested in knowing the level, at a given frequency, in a band of one cycle per second band width from which we can readily compute the level in any band width.

Equipment used in measuring noise levels is often equipped with filters to isolate certain frequency bands. For some kinds of analyses, filters having band widths of one octave, ½ octave and ⅓ octave are often used. These filters operating between frequencies f_1 and f_2 have the following characteristics, where C is the band width and f_g is the center frequency (geometric mean).

One octave band filter
$$C = f_2 - f_1 = 0.707 \, f_g$$
$$f_g^2 = f_1 f_2, \; f_1 = 0.707 \, f_g$$
$$f_2 = 2f_1, \; f_2 = 1.414 \, f_g$$

½ octave band filter
$$C = f_2 - f_1 = 0.348 \, f_g$$
$$f_g^2 = f_1 f_2, \; f_1 = 0.841 \, f_g$$
$$f_2 = (2)^{\frac{1}{2}} f_1, \; f_2 = 1.189 \, f_g$$

⅓ octave band filter
$$C = f_2 - f_1 = 0.231 \, f_g$$
$$f_g^2 = f_1 f_2, \; f_1 = 0.890 \, f_g$$
$$f_2 = (2)^{\frac{1}{3}} f_1, \; f_2 = 1.121 \, f_g.$$

We often have noise contributed by two or more sources. If the noise is contributed by two sources of equal intensity and one is eliminated, the intensity level will be reduced by three dB (see Fig. 2.1). If the two sources are not of equal intensity, eliminating the one of lower intensity will reduce the total intensity level by an amount less than three dB.

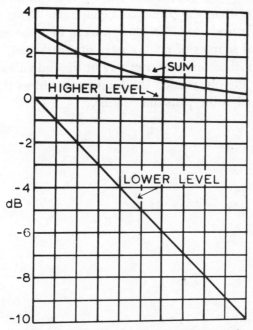

Fig. 5.1. Chart giving the result of addition of two noise signals in decibels.

Figure 5.1 is a chart showing the increase in intensity level produced when two noise levels are added. To use this chart, find the ratio of the lower signal level to the higher signal level in decibels. Locate this value on the ordinate scale (vertical axis) of the bottom curve. The abscissa (horizontal axis) corresponding to this ordinate on the bottom curve is the abscissa corresponding to the ordinate on the top curve which determines the number of decibels by which the higher level signal will be increased when the lower level signal is added to it. For example, if we add a signal six dB lower than the principal signal, the sum of the two signals will be about 0.8 dB higher than the principal signal alone. This is an important consideration in the problem of noise reduction. If several sources of noise contribute to the total noise in a system, the elimination of any of the individual sources will not greatly reduce the total noise unless the source eliminated is the principal source.

6.
Transducers

6.1. *Introduction.*

A transducer is any device capable of converting energy of one kind into energy of another kind. An automobile engine is a transducer that converts the chemical energy of the fuel and oxygen into mechanical energy.

In acoustics, we use transducers to convert sound energy to electrical energy or to convert electrical energy, hydraulic energy or chemical energy to acoustic energy. Transducers used in air to convert acoustic energy to electrical energy are called *microphones,* if they are used in water they are called *hydrophones.* If they are used to convert another form of energy to acoustic energy, they are called *sound projectors* or, if no confusion would result, *projectors.*

Microphones may respond to sound pressure or sound particle velocity, while hydrophones nearly always are designed to respond to the sound pressure. Another kind of transducer, designed to measure accelerations in vibrators, responds to the acceleration of the vibrator and is called an *accelerometer.*

Many transducers that can be used to convert acoustic energy to electrical energy will also operate to convert electrical energy to acoustic energy. Such transducers are called *reciprocal transducers.*

6.2. *Transducer Elements.*

There are several ways in which it is possible to convert the sound pressure or the particle velocity of a sound wave to an electric signal. We can classify these as *piezoelectric, electrostrictive, electrodynamic, magnetostrictive* and *electrostatic*.

In 1880 the Curie brothers demonstrated that certain crystals, when compressed or expanded in particular directions, develop positive and negative charges on certain portions of their surfaces; the charges are directly proportional to the pressure and they change sign when the compression changes to expansion. The charges disappear when the pressure is released. The Curies demonstrated later that when charges are applied to the surfaces of such crystals, a strain, which is proportional to the charge and changes sign with it, is developed in the crystal.

When any insulator is subjected to an electric field, it is deformed. This deformation is called electrostriction. The deformation is independent of the direction of the electric field and is proportional to the square of the electric field strength and it therefore does not change sign with it. The electrostrictive effect is common to all materials and is so much smaller than the piezoelectric effect in most instances that it can ordinarily be ignored even though it is always present when the piezoelectric effect is manifested. An important exception occurs with certain ceramics such as barium titanate and lead zirconate titanate in which the electrostrictive effect is large in comparison with the piezoelectric effect. If these materials are polarized so that they are permanently subjected to a high value of electric field, they behave in a manner similar to that of the piezoelectric crystals. The ceramic materials

have an important advantage in that they can be molded
to any desired size and shape, while crystal elements must
be cut from larger single crystals of the material.

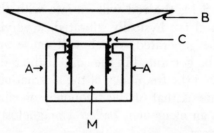

Fig. 6.1. Cross section of a common loudspeaker. *M* is the
magnet, *C* is the coil, *AA* is the iron armature and
B is the paper cone.

If a coil of wire is arranged so that it can be moved in a
magnetic field, a voltage will be generated in the coil that
is proportional to the strength of the field, the velocity of
motion of the coil and the length of wire in the coil that
cuts the field. Figure 6.1 shows schematically a cross
section of a common arrangement for such a system. The
portion *M* is a magnet, *AA* is an iron armature and the
coil *C* is arranged in the flux gap of the system. If the coil
is moved up and down, it will cut the magnetic flux in the
gap and an alternating electric current will be generated
in the coil. If, on the other hand, an alternating current
is supplied to the coil, it will move up and down with the
same frequency as the current. The ordinary loud speaker
is a simple example of such a system.

If a rod of nickel or some of its alloys is surrounded by
a coil of wire carrying an electric current the rod will
change in length. Whether the rod lengthens or shortens
is determined by the material of the rod, but it is inde-

pendent of the direction of the current. If an alternating current of frequency f is supplied to the coil, the frequency of vibration of the rod will be equal to $2f$. However, if a magnet is incorporated with the rod to produce a polarizing field, the rod will vibrate with the frequency f. If the rod is physically vibrated, a varying magnetic flux will be generated, which will cause an alternating current to be generated in the coil. If the rod is magnetically polarized the frequency of the alternating current will be the same as that of the frequency of vibration of the rod. Such an element is called a magnetostrictive transducer element.

The electrostatic transducer element utilizes the fact that the electrical capacity of an electric condenser depends on the spacing between the plates, and for a given electric charge on the condenser the voltage varies inversely with the capacity of the condenser.

6.3.   *Microphones.*

Microphones may be pressure-actuated or pressure-gradient actuated. Figure 6.2 shows the basic configuration of a pressure-actuated microphone. The diaphragm A is acted on by the pressure variations in the incident

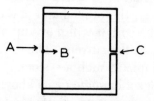

Fig. 6.2.   Pressure-actuated microphone. A is the diaphragm, B is the mechanical connection to the transducer element and C is a small air leak.

sound wave and will therefore be alternately deflected to the right and left as the pressure at the front varies above and below the ambient pressure. The diaphragm is connected to the transducing element by the mechanical connection *B*, and *C* is a small air leak that permits the space in the cavity behind the diaphragm to assume the same static pressure as the ambient pressure in front of the diaphragm.

A pressure-gradient microphone responds to the gradient of the pressure, which is related to the particle velocity. Such a microphone has a diaphragm connected to the transducing element in a manner similar to that of the pressure-actuated microphone, but it is so arranged that both sides of the diaphragm are exposed to the sound signal so the response of the daphragm is determined by the difference in pressure on the two sides. Figure 6.3 shows

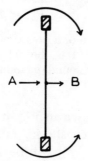

Fig. 6.3.   Pressure-gradient microphone. *A* is the diaphragm and *B* is the mechanical connection to the transducing element.

the physical arrangement of the diaphragm in the pressure-gradient microphone. An important feature of this microphone is that its response to sound incident at the edge of the diaphragm is zero, since there would then be no pressure gradient. This is a desirable feature when one wishes to have a miximum response of the microphone to sound from a specific direction and to discriminate against sound from other directions. Figure 6.4 shows the pattern of the response of such a microphone. The directions of maximum response are along the line through the center of the microphone, perpendicular to the diaphragm.

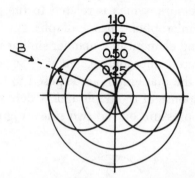

Fig. 6.4.   Response pattern of a pressure-gradient microphone
plotted in dB. *A* is the relative response of the
microphone for sound incident from the direction
B.

If the microphone shown in Fig. 6.2 is made with an opening containing an acoustic impedance in the back in place of the small air leak, it will be a combination pressure-and pressure-gradient microphone. Such a microphone is commonly called a *cardioid* microphone because of the shape of the response pattern. Figure 6.5 shows a response curve for a cardioid microphone with the acoustic impedance of the opening so chosen that there is a null in the response pattern at the rear of the microphone.

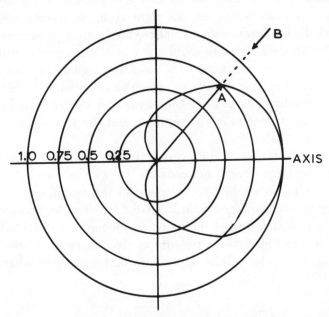

Fig. 6.5. Response pattern of a cardioid microphone plotted in dB. A is the relative response of the microphone for sound incident from the direction B.

6.4. *Methods of Incorporating Transducing Elements in Transducers.*

The method of incorporating the transducing element in a transducer and the design of the transducing element will depend on its particular application.

Piezoelectric and ceramic elements are extensively used in microphones because they are relatively inexpensive and rugged. Crystals commonly used for microphones are Rochelle salt, ammonium dihydrogen phosphate and lithium sulphate in addition to the ceramic, barium ti-tanate. The ceramics are the most commonly used ma-terials in hydrophones and underwater sound projectors.

The stresses that the diaphragm of a microphone can exert on the element are very small for a given intensity of sound, because of the low acoustic impedance of air. The electrical response of the element is determined by the strain produced in the element, and the strain is pro-portional to the stress.

The orientation of the crystal axes in the element cut from the large crystal depends on the way in which the stress is to be applied to the plate in the transducer. A shear plate is arranged as in Fig. 6.6. A metal electrode is applied to each side of the plate, and when it is deflected —as at *a* in Fig. 6.6—the polarity of the charge developed on the electrodes will be opposite to that developed when it is deflected as in *b*.

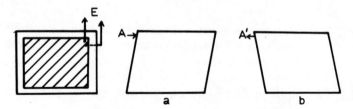

Fig. 6.6.   Shear plate element. The electrode connections are indicated at *E*. *A* and *A'* indicate the shear forces on the plate in conditions *a* and *b*.

If a rectangular plate is cut from a shear plate so that its long axis is parallel to one diagonal of the shear plate, it is called an expander plate or an expander bar. If the stress is applied parallel to the long axis of the rec-tangle, electric charges will develop on electrodes placed on the two sides of the plate. Expander plates may be used as length expanders or thickness expanders. A thickness expander has the stress applied over a large area side with the opposite side rigidly mounted.

An important disadvantage of the expander plates and the shear plates for use in microphones is the large mechanical impedance of the crystals so that the strains developed by a given intensity of sound in air, which has a very low acoustic impedance, will be very small. To reduce the mechanical impedance of the crystal element without lowering the voltage output, two plates can be combined to produce a Bimorph. (Bimorph is a registered trademark of the Brush Electronics Company, Cleveland, Ohio.) The two plates forming the Bimorph are so combined that when a force is applied to bend it, a tensile stress is generated in one plate and a compressive stress is generated in the other. The electrodes may be applied to a Bimorph in two ways to form either a series or a parallel arrangement of the plates. If a series connection is desired the terminals are connected to the two outer foils. For a parallel connection, the two outer foils are connected together to form one terminal and the foil between the crystals forms the other terminal. The choice of series or parallel connection is determined by the electrical characteristics of the equipment to which it is connected.

Crystal and ceramic elements are commonly used as the sensitive elements in hydrophones and underwater sound projectors. The elements are usually coupled to the water medium with a rubber window between the crystal or ceramic element and the water. A special rubber having the same acoustic impedance as the water is used for the window and a film of castor oil is used between the element and the rubber in order to effectively couple them and eliminate any entrapped air. The ambient pressure varies greatly with depth in water so hydrophones and underwater sound projectors may need to be designed to withstand several thousand pounds per square inch. One form of ceramic element used quite extensively in underwater sound transducers is a ceramic ring. These rings may be molded in various sizes depending on the frequencies at which they are to be operated. The electrodes are applied to the inside and outside surfaces of the rings and the rings may be used as either sound projectors or receivers. When such rings are used as transducers the radial mode of vibration is used. The crystal and ceramic transducers are reciprocal transducers but many of the units designed as microphones would have very little power handling capability as projectors.

The most common application of the electrodynamic principle in transducers is that used in the ordinary loudspeaker. A coil is wound on a cylindrical tube suspended from a cone-shaped diaphragm in a radial magnetic field. When used as a microphone, the pressure variations on the cone due to the sound signal cause the coil to move in the magnetic field generating a voltage in the coil which varies as the sound pressure. The speaker is more commonly used as a sound projector. The coil will be forced to oscillate in the magnetic field at the frequency of an alternating current supplied to it.

Another common form of electrodynamic microphone is shown in Fig. 6.7. The coil with terminals A and B is wound on the two legs which are the pole pieces of the magnet M, and the magnetic circuit is completed through the steel diaphragm D, leaving two air gaps between the pole pieces and the diaphragm. When sound waves impinge on the steel diaphragm, vibrations are forced, vary-

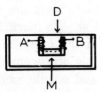

Fig. 6.7.   Electrodynamic microphone. *A* and *B* are the terminals of the coil, *M* is the magnet and *D* is the
diaphragm.

ing the air gaps in the magnetic circuit and thus causing
the magnetic flux in the coils to vary so a voltage having
the same frequency as the sound pressure is generated in
the coil.

A form of electrodynamic transducer often used as a
phonograph pickup is shown in Fig. 6.8. The needle *N*

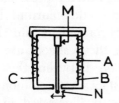

Fig. 6.8.   Electrodynamic phonograph pickup. The needle,
N, is at the end of a thin spring of magnetic material. *M* is the magnet and *C* and *B* are the pole
pieces.

follows the wavy groove in the record and moves back
and forth as indicated by the arrows. When the armature *A* moves to the right, the magnetic flux from the
magnet *M* will be increased through the pole piece *B* and
decreased through the pole piece *C*. These changes in
flux will generate voltages in the coils which are so con-

nected that the voltages in the two coils will add. When
the armature moves to the left the flux at pole piece A
will decrease and that at pole piece B will increase. This
will cause a voltage of the opposite polarity to be generated in the coils. This arrangement has an advantage
in that the armature and needle can be made with very
small mass and the bending modulus of the armature
supplies the elastic restoring force.

The most common application of magnetostrictive elements is in underwater sound transducers. The magnetostrictive materials are electrical conductors, and if a material which is an electrical conductor, such as a rod of
nickel, is placed in the vicinity of a coil carrying an alternating current, a voltage will be induced in the metal
causing currents—called eddy currents—to flow. These
eddy currents wil cause heat to be generated in the magnetostrictive material, which represents an energy loss. In
order to minimize the eddy current losses, the magnetostrictive material is formed in thin sheets, and laminations stamped from these sheets are consolidated into
stacks with insulation between the laminations. Two
forms of laminated stacks are commony used. One form,
called hairpin stacks, is illustrated in Fig. 6.9. Coils C
are placed on the narrow portions of the legs and a magnet M is placed between the legs to provide polarization
and complete the magnetic circuit. These stacks are
usually made so that their length is equal to one-half a
wavelength in the metal for the frequency at which the
stack will be operated and they are usually operated at
this resonant frequency.

The ring stack is another form of magnetostrictive laminated stack. Rings of the laminations are cut and they
are assembled into a stack like that shown in Fig. 6.10

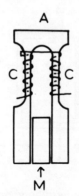

Fig. 6.9. Magnetostrictive hairpin transducer element. A is the radiating face of the element, CC are the coils and M is the polarizing magnet.

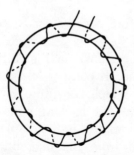

Fig. 6.10. Magnetostrictive ring stack transducer.

and the coil is wound toroidally on the stack. The stack vibrates radially—that is, it alternately increases and decreases its radius. The ring stack may be polarized by use of a supplementary coil carrying a direct current or the magnetostrictive material may be a special alloy such as permendur, which is capable of permanent magnetization. The magnetostrictive transducers are reciprocal trans-

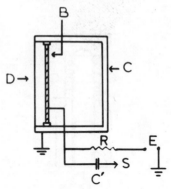

Fig. 6.11. Condenser microphone. *D* is the diaphragm, *B*
is the backing plate, *C* is the microphone shell, *R*
is a high resistance resistor in the backing plate
charging circuit, *C'* is an electrical condenser and
S is the signal terminal.

ducers and therefore may be used either as receivers or
sound projectors.

The microphone shown in Fig. 6.11 is called a condenser microphone. It consists of a diaphragm *D* with a
backing plate *B*, which constitute the two plates of an
electric condenser. The backing plate is maintained at a
high D.C. potential relative to the diaphragm. The backing
plate is insulated from the microphone case and diaphragm, which are maintained at ground potential. The
electrical capacity of the condenser is proportional to the
area of the diaphragm and inversely proportional to the
spacing between the diaphragm and the backing plate.
With a fixed electric charge *Q* on the backing plate, the
potential difference *V* between the diaphragm and
the backing plate will be $V = Q/C$, where *C*
is the electrical capacity of the system. If the external
pressure on the diaphragm is increased slightly, it will

be deflected toward the backing plate. The spacing will be decreased, which will increase the electrical capacity. Since the quantity of charge will not change, the increase in capacity will cause a decrease in the voltage at the backing plate. Conversely, when the external pressure is decreased, the diaphragm will be deflected away from the backing plate causing the capacity to decrease, which will result in an increase in the backing plate voltage. The charge on the backing plate is maintained from a fixed power supply with a high resistance in the circuit so that the charge will remain essentially constant when the diaphragm is actuated by the sound pressure variations in a sound signal. The voltage variation on the backing plate will correspond to the pressure variations—due to the sound signal—on the diaphragm. The chamber back of the backing plate is maintained at the ambient pressure of the atmosphere by means of a slow leak in the back of the microphone case. Provision is also made for leakage of air from the space between the diaphragm and backing plate and the space back of the backing plate. This serves to maintain the ambient pressure in the space between the diaphragm and the backing plate and it is also designed to serve as a means of damping the vibration of the diaphragm. A perforated cover is placed over the diaphragm to protect it from mechanical damage. In Fig. 6.11, E is the source of D.C. voltage for the backing plate and R is the high resistance in this circuit. The D.C. voltage is isolated from the A.C. signal voltage S by means of the condenser C'.

Condenser microphones are precision instruments. They are expensive to manufacture but they are used almost exclusively in sound measuring equipment in air acoustics.

Transducers designed to work at some specific fre-

quency are often designed to operate at resonance. The
response of the transducer at resonance will, of course,
be greater than at some frequency off resonance. If the
transducer is to be operated over a large band of frequencies it will be designed so that the resonant frequency is
either higher or lower than any of the frequencies of interest. The ordinary loudspeaker is designed to have its
resonance at the low end of the operating frequency
range. In this instance the resonant frequency is included
in the operating range but its effect is minimized by the
use of proper damping and design of the enclosure in
which it is mounted.

6.5.   *Transducer Arrays and Directivity.*

The transducers which have been so far considered are
small compared to the wavelength of the sound transmitted or received. Such transducers are generally omnidirectional—that is, they have essentially the same response in all directions. The pressure gradient and cardioid microphones are, of course, exceptions. If, however,
the dimensions of the transducer are large compared to
the wavelength of the sound, the transducer will have
directional characteristics. There will be one direction,
called the acoustic axis of the transducer, in which its
sensitivity as a receiver or its effectiveness as a projector
will be a maximum and its response will decrease as the
angle from the acoustic axis is increased. The ratio of the
square of the voltage produced in a receiver in response
to sound waves arriving on the acoustic axis to the average square of the voltage produced if sound waves having
the same frequency and mean square pressure were arriving at the transducer simultaneously from all directions
is called the *directivity factor* of the transducer as a re-

ceiver. The directivity factor of a sound projector is equal to the ratio of the intensity of the transmitted sound at a remote point on the acoustic axis to the average intensity of the sound transmitted through the surface of a sphere passing through the remote point, the center of which is at the transducer. If the transducer is a reciprocal transducer, its directivity factor as a sound projector will be the same as its directivity factor as a sound receiver. The directivity factor of a piston is equal to k^2R^2 where R is the radius and $k = 2\pi/\lambda$.

We can convert the directivity factor of a transducer to dB by use of Fig. 2.1. When the directivity is expressed in dB, it is called the *directivity index*. A transducer having a directivity factor of 2 will therefore have a directivity index of 3 dB.

In order to produce a transducer with an appreciable directivity index, its linear dimensions must be large compared to a wavelength of the sound. In order to achieve this, it is generally necessary to construct the transducer with an array of elements since it is not feasible to construct a single transducing element of sufficient size. The most important application of arrays of elements is in underwater acoustics.

If a line array of elements is used, the array will be directional in the plane of the line but it will be omnidirectional in the plane perpendicular to the line. An interesting application of a line array is sometimes used in sound reinforcing systems in cathedrals. The ceiling in a cathedral is very high but the congregation is seated in a horizontal plane near the floor. The sound reinforcement is desired only in the plane occupied by the congregation, so vertical arrays of speakers are used to produce a fan shaped pattern of sound in the plane occupied by the congregation. Antenna arrays, which are the transducers for radio waves, behave in the same manner as acoustic transducer arrays, and since television receiver antennas are located on the surface of the earth, the television transmitter antennas are tall vertical arrays to produce a pattern which is highly directional in the vertical plane but omnidirectional in the horizontal plane. If the directivity index of a television transmitter antenna is 10 dB, the intensity of signal at the receiver antenna will be 10 dB higher than it would be if the transmitter antenna were omnidirectional. To produce the same intensity at the receiver with an omnidirectional antenna would require a power output from the transmitter 10 times as great.

Directional arrays used as receivers in underwater acoustics help in discriminating against ambient noise in the ocean. The ambient noise signal measured on a receiving transducer having a directivity index of 10 dB will be 10 dB lower than that measured on an omnidirectional transducer.

6.6. *Non Reciprocal Transducers.*

Non reciprocal transducers have a greater application in underwater acoustics than in air acoustics. The most important of these is explosive sources. Explosive sources range from small dynamite caps to several pounds of TNT. By means of large explosive charges, signals can be transmitted over ranges of several thousand miles.

Another important source of sound often used under water is the underwater spark. The spark is produced by discharging a large bank of high voltage condensers across a spark gap to produce a very intense sound of very short duration.

7.
Sound Measurement and Analysis

7.1. *Introduction.*

In order to measure or analyze sound it is necessary to detect the sound waves with a calibrated microphone or hydrophone. The sound signal intercepted by the transducer causes an electrical signal to be generated which must be amplified by means of an electronic amplifier in order to develop sufficient power to actuate a device such as a meter.

The electrical signal generated in the receiving transducer contains all frequencies present in the sound unless there are sound frequencies present to which the transducer does not respond. If measurements or analyses are to be carried out over a frequency range greater than that to which a given transducer will respond, it is possible to use more than one transducer to cover the total desired frequency range. Each transducer is connected to its own electronic system. In making measurements in the sonic range in air, the receiving transducer most commonly employed is a condenser microphone. A good condenser microphone responds to a frequency range about the same as that of the human ear.

The state of the art in frequency control of electronic amplifiers is highly developed. Electrical filters can be designed to pass electrical signals of any desired frequency and band width, so an electrical system can be designed to pass the desired band or to compensate for any uneven frequency response characteristics of the receiving transducer.

7.2. *Measurement of Sound Pressure Levels.*

Sound is usually measured by measuring the sound pressure. We may be interested in the total sound pressure or in the sound pressure in some limited band of frequency. The basic instrumentation for measuring

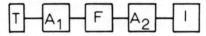

Fig. 7.1. Basic instrumentation for measurement of sound pressure levels.

sound pressures is shown schematically in Fig. 7.1, where T is the receiving transducer, A_1 is a preamplifier, F is an electrical signal filter, A_2 is an amplifier, and I is the instrument for recording or indicating the sound pressure. In many instances, the electrical signal generated by the receiving transducer is at such a low power level that the preamplifier must be especially designed to have a very low *input noise* level and, in some instances, it is incorporated with the receiving transducer. When we speak of the input noise level of an amplifier, we do not mean noise in the sense of acoustic noise, but rather, we are referring to unwanted voltage variations that appear in a circuit element because of stray signals or because of thermal energy existing in the resistance of the input circuit of A_1. The *thermal noise* is a function only of the temperature, the

resistance of the input circuit and the characteristics of
the first amplifier element in the amplifier A_1. The noise
caused by stray electric or magnetic fields is picked up by
the cable connecting the transducer with A_1 and is minimized by incorporating A_1 with the transducer. Recently,
special transistors have been developed, called field effect
transistors, which have especially low input noise levels.
When very low level sound fields are to be measured, field
⬚ r the preamplifier.
⬚ nounts of power for
⬚ ry can easily be in-

⬚ ng the sound gen-
⬚ certain that other
⬚ signal of sufficient
⬚ urement. If the re-
⬚ i. e., has the same
⬚ , great care must be
⬚ ambient noise does
⬚ However, if the re-
⬚ ctional, it will dis-
⬚ uch transducers are
⬚ than in air acoustics
⬚ must have linear di-
⬚ the wavelength in
⬚ vity; and the wave-
⬚ quency is about five
⬚ because the velocity
⬚ s as great as it is in

⬚ are used, great care
⬚ t the source of the
⬚ of maximum sensi-

tivity of the transducer because of the rather high variation of the sensitivity of such a transducer with the angle
of incidence of the sound.

7.3. *The Sound Level Meter.*

In order to determine the noise levels in factories, communities and any other areas in which noise is either a
hazard or a nuisance it is necessary to have an easily portable meter to measure the noise. The hazard or nuisance
effect of noise depends on the characteristics of the human ears. At low noise levels, the sensitivity of our ears
to low frequencies and very high frequencies is less relative to the sensitivity in the midrange than it is at high
noise levels. Therefore, as the noise level increases, the
relative sensitivity of our ears in these extreme ranges increases. The sound level meter must, therefore, reflect
this characteristic in order to provide measurement consistent with the characteristics of our ears over all ranges
of sound level.

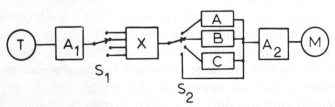

Fig. 7.2. Block diagram of a typical sound level meter.

Figure 7.2 shows a block diagram of a typical sound
level meter, where T is the receiving transducer—usually
a condenser microphone—A_1 is the preamplifier; X is a
series of calibrated attenuators or range selectors controlled by the switch S_1; A, B and C are three weighting

networks controlled by the switch S_2; A_2 is an amplifier and M is the meter on which the sound pressure level is read.

The sound level meter measures the level in dB relative to 0.0002 microbar. The attenuators in X are calibrated in dB so the level measured on the meter added to the reading of the position of S_1 gives the level of the sound in dB.

If S_2 is in the position that bypasses the weighting networks the meter will read the true sound pressure level. When the weighting networks are used, the meter reads a level corresponding to the response of an average human ear, and since this is not a true sound pressure level we call it the *sound level*. At low sound levels, where the ear has its minimum sensitivity for low and very high frequencies, the A scale is used. At intermediate sound levels, the B scale is used and at high sound levels the C scale is used. Figure 7.3 shows the relative response of such a meter with each of the three weighting networks.

It is necessary to use considerable care in making sound

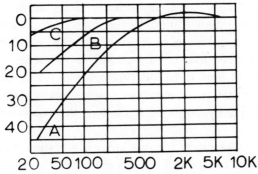

Fig. 7.3. Response curves for the A, B and C scales of a sound level meter.

level measurements with the sound level meter because the presence of the observer may make an appreciable error in the reading. The observer should move around while making the measurement in order to be certain that he makes a representative measurement of the sound level. When measurements of the sound level are made inside of buildings, great care must be exercised to be certain that representative readings are obtained because reflections of the sound within the structure may play an important role in determining the sound level at any position.

7.4. *Measurement of Sound Pressure Levels as a Function of Frequency.*

Measurements with the sound level meter tell us the value of the total sound level but give us no information about the quality or spectral distribution of the sound. In order to obtain the necessary information to be able to reduce the noise emitted by some mechanism, it is necessary to know how the noise varies with frequency over the sonic range. A common means of accomplishing this is to use a system similar to the sound level meter with special electrical filters that can be inserted in place of the weighting networks of the sound level meter. Such a system is not limited to measurements in the sonic region and we can use it for measurements of underwater sound pressures if a hydrophone is used as the receiving transducer.

The kinds of filters for use in such a system have been standardized as one octave, ½ octave and ⅓ octave band widths. The one-octave filters pass a frequency band one octave wide and the ½ and ⅓ octave filters pass respectively

½ octave and ⅓ octave bands. The center frequencies for sets of such filters have been standardized and the center frequencies for the one octave and the ⅓ octave filter sets are given in Tables 7.1 and 7.2.

Table 7.1
1/1 Octave band center frequencies.

f_0 c.p.s.	f_0 c.p.s	f_0 c.p.s.	f_0 c.p.s
16	125	1000	8000
31.5	250	2000	16000
63.	500	4000	31500

Table 7.2
1/3 Octave band center frequencies.

f_0 c.p.s.	f_0 c.p.s	f_0 c.p.s.	f_0 c.p.s
12.5	125	1250	12500
16	160	1600	16000
20	200	2000	20000
25	250	2500	25000
31.5	315	3150	31500
40	400	4000	40000
50	500	5000	
63	630	6300	
80	800	8000	
100	1000	10000	

A frequency analyzer consists of a set of standard filters with a receiving transducer and suitable amplifiers and switches with a level recorder used in place of a meter. The level recorder operates the switch, which selectively connects the filters into the circuit, one after the other. An analyzer manufactured by the Brüel and Kjaer company can scan the frequency range from 10 c.p.s. to 45,000 c.p.s. in 18 seconds. The sound level recorder plots the sound pressure level in dB as a function of frequency, as a pen and ink plot on paper. Such spectrum analyses are usually not made directly at the source of the sound because such instruments are not readily portable. The sound signal is detected with the receiving transducer, amplified and recorded on magnetic tape. More than one receiving transducer may be used to cover the required spectral range and special filter networks may be used to correct for any lack of uniform frequency response of the receiving transducers and the magnetic recording and playback systems. A loop of the magnetic tape recorded at the time of interest is formed from a section of the total tape and this loop is run around and around over the tape playback head and the signal from the tape is amplified and fed into the analyzer.

7.5. *Narrow Band Spectrum Analysis.*

In some instances a more detailed spectrum analysis of a sound than that provided by the ⅓ octave filters is desired. By means of a narrow band spectrum analysis it is often possible to isolate discrete sources of noise such as the meshing of teeth in gears or a singing element on an airplane structure. We can use the same magnetic tape recording to do a narrow band spectrum analysis as we used to do a ⅓ octave band analysis.

By reference to Table 7.2 we can see that it is necessary to have 35 filters to cover the range from 12.5 c.p.s. to 40,000 c.p.s. in ⅓ octave bands. To perform a narrow band spectrum analysis, filters of five c.p.s. bandwidth are usually used below 10,000 c.p.s. and 50 c.p.s. band width filters are usually used for frequencies above 10,000 c.p.s. If separate filters were used 2000 separate five c.p.s. filters would be required to cover the frequency range below 10,000 c.p.s. The five c.p.s. filters are quite expensive so it is obvious that such an arrangement would be highly impractical. This problem is solved by means of a system that utilizes a single high precision filter to cover the entire range below 10,000 c.p.s.

In order to understand how such a system operates, let us consider what we would hear if we were tuning some musical instrument to the a (440 c.p.s.) key on a piano. As we approach exact tuning, we will hear beats between the two tones. The frequency of the beats is equal to the difference between the frequencies of the two tones, so if the piano tone is at 440 c.p.s. and the frequency of the tone of the other instrument is at 438 or 442 c.p.s. we will hear a beat note of two c.p.s. We can generate such a beat frequency in an electrical circuit by properly mixing two electrical signals. If we mix an electrical signal of frequency f_1 with another signal of frequency f_2 there will be the four resulting signals—f_1, f_2, $f_1 + f_2$ and $f_1 - f_2$. By means of a suitable electrical filter we can eliminate the signals f_1, f_2 and $f_1 + f_2$, leaving only $f_1 - f_2$. A narrow-band spectrum analysis system has a tunable oscillator to generate the signal of frequency f_1 and this oscillator is so designed that f_1 can be continuously varied over some range of higher frequency than any of the frequencies in the sample to be analyzed. If we wish to analyze a sample over the frequency range from zero to 10,000 c.p.s., the tunable oscillator will be tuned over a range of frequency higher than 10,000 c.p.s. A narrow band pass filter having a fixed center frequency f_0 with a band pass of five c.p.s. will then be used to pass the beat frequency $f_1 - f_2$. Figure 7.4 is a schematic diagram of a typical narrow band spectrum analysis system.

In this arrangement, T is a tape playback unit that plays back the signal from a loop of magnetic tape cut from a

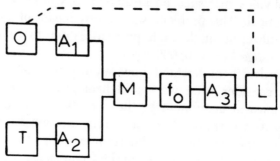

Fig. 7.4.   Schematic diagram of a typical narrow band sound
analysis system.

tape recording of the sound signal recorded on a magnetic
tape recorder. The signal from the magnetic tape playback head is amplified in the amplifier A_2, which contains
filter networks to correct for any nonuniformities in the
playback head response and the signal is fed to the electrical signal mixer M. A precision tunable oscillator O
generates a signal of·frequency f_1, which is amplified in
the amplifier A_1 and also fed to the mixer. The output of
the mixer will contain the frequencies f_1, f_2, $f_1 + f_2$ and
$f_1 - f_2$. The signal that passes the precision narrow band
filter f_0 will be the beat signal corresponding to the particular frequency, f_2, recorded on the tape for which $f_0 =
f_1 - f_2$. This signal is amplified in the amplifier A_3 and
recorded on the pen and ink level recorder L. In order to
record a complete spectrum, an electric motor drives the
paper feed on the level recorder and the paper feed is
connected by means of a shaft, indicated by the dashed
line in Fig. 7.4, to the tuning dial on the oscillator, O. The
procedure, then, is to set the oscillator dial to the proper
frequency at one end of the frequency range, and while
the tape loop is run around and around over the playback

head the motor drives the oscillator through the desired range of frequency and drives the recording paper in synchronism with it so that the spectrum showing the level in dB as a function of frequency is automatically plotted. Figures 7.5 and 7.6 show the spectrum plots made from the same tape recording with a ⅓ octave band spectrum analyzer and a narrow band spectrum analyzer. These curves illustrate the advantage of the narrow band spectrum analyzer in delineating narrow band sources of sound.

The disadvantage of the narrow band spectrum analyzer is that it takes considerably more time to do the analysis because the recording must be done slowly enough to allow the signal in the narrow band filter to build up to its full value at each frequency. If the detailed

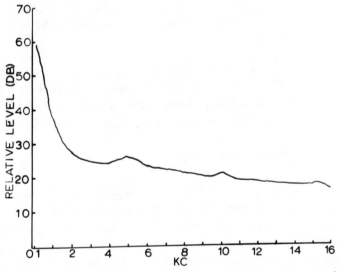

Fig. 7.5.   Typical ⅓ octave band analysis of a recording in the sonic region.

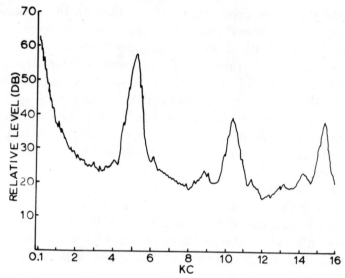

Fig. 7.6. Narrow band analysis of the same recording as that used for the ⅓ octave band analysis shown in Fig. 7.5.

frequency analysis indicated in Fig. 7.6 is not required, the ⅓ octave or the ¹⁄₁ octave analyzer can provide the information in a much shorter time and this can be important when large numbers of analyses are required.

7.6. *The Sound Spectrograph.*

The methods of analysis described in the last section are applicable only for sounds that are continuous in time for at least several seconds. The magnetic tape loops used for octave band or narrow band spectrum analysis represent recordings made over a two-to ten-second interval and the analysis represents the average characteristics of the sound over that interval of time.

There are many sounds—such as voice sounds, heart sounds and sounds emitted by animals—for which an analysis giving average characteristics over a period of time is relatively meaningless. What is needed is an analysis showing both frequency and the relative intensities of the various frequencies as a function of time. The sound spectrograph is designed to perform such a three dimensional analysis.

Figure 7.7 shows one of the commercially available sound spectrographs. The drum at the top of the instru-

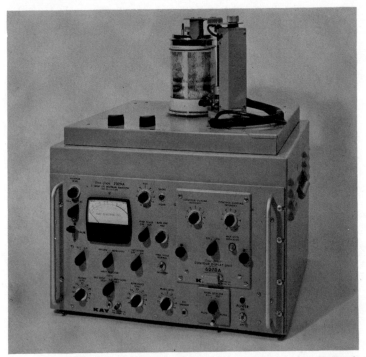

Fig. 7.7. Photograph of a sound spectrograph. Photograph courtesy of Kay Electric Co.

ment carries the recording paper on which the analysis record is produced. The recording paper is impregnated with a chemical that darkens when an electric current passes through it and the amount of darkening is a measure of the current density. The stylus is a fine wire that rubs on the drum as it rotates and an electric current determined by the signal level flows from the wire, through the sensitive paper to the metal drum.

The axis of the drum is locked to the axis of a turntable within the instrument so that they will rotate in synchronism. The rim of the turntable is coated with a magnetic material on which a signal can be recorded in the same manner as recordings are made on magnetic tape. The signal to be analyzed is recorded on the rim of the turntable and the time duration of the recorded signal is equal to the time required for one revolution of the drum and turntable.

The frequency analysis of the signal is carried out in a manner similar to that of the narrow band spectrum analyzer, i. e., a local oscillator generates a high frequency signal which, when mixed with the playback signal from a magnetic transducer at the rim of the drum, will produce the two frequencies $f_0 + f$ and $f_0 - f$, where f_0 is the frequency of the oscillator and f is the frequency of a signal received from the playback transducer. This signal goes to a narrow band filter and if the signal f_0-f is at the frequency of the narrow band filter it will pass the filter and be amplified. This amplified signal will be fed to the stylus and its magnitude will determine the amount of electric current that flows from the stylus through the sensitive paper to the drum.

The stylus is moved from the bottom of the paper to the top by means of a screw and the shaft of the screw is mechanically coupled to the frequency control of the local oscillator. When the stylus is at the bottom of the paper, the local oscillator will generate a frequency equal to the center frequency of the narrow band filter so the record at the bottom of the paper will correspond to a signal frequency f equal to zero. As the drum rotates, the screw will move the stylus up a distance approximately equal to its width during one revolution and the local oscillator frequency will be increased by a small increment. In this way the recorded signal is scanned over and over at increasing frequencies so that relative times can be observed around the circumference of the drum and increasing frequencies observed from bottom to top of the drum. When the paper is removed from the drum and laid flat on a table, relative times will increase from left to right and frequencies will increase from the bottom to the top of the paper. If several pure tones are recorded on the turntable, the record will be clear except for horizontal lines corresponding to those frequencies on the paper and if they are recorded for only part of the revolution, they will appear on the time axis corresponding to the times at which they were recorded.

The instrument shown in Fig. 7.7 has a number of speeds at which the signal is recorded in order to provide different frequency ranges for the width of the paper. The range of frequencies that can be analyzed with this instrument is from zero to 16 kc. When the range to be covered is eight kc, the signal will be recorded at 25 revolutions per minute and played back for analysis at 300 revolutions per minute and the length of signal to be analyzed will be 2.4 seconds.

The dynamic range of the paper is 10 dB but by use of special automatic gain control circuits a signal intensity range of 40 dB can be compressed to the 10 dB dynamic range of the paper. Figure 7.8 is a recording showing density and frequency as a function of time.

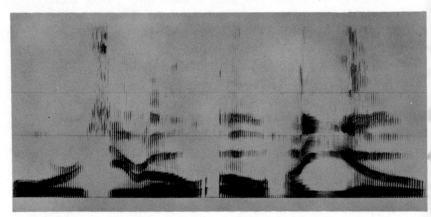

Fig. 7.8.   Record made with a sound spectrograph using varying density to indicate varying intensity. Courtesy
of Kay Electric Co.

By means of accessory equipment this sound spectrograph can be adapted to plot contours at the various frequencies so that the contour lines represent 6 dB intervals
of intensity. The intensities at each frequency and time
can be determined by counting the contour lines rather
than by estimating the density on the record. Figure 7.9
shows an analysis of a voice recording by this method
using the spoken words DID YOU THANK HIM? Figure 7.10
shows a heartbeat recording using the contour recording
method. Note the indication of heart murmur on the recording.

Some observers claim that the sound spectrogram of a
voice is as characteristic of an individual as his finger-
prints. For this reason, sound spectrograms of voices are
sometimes called voiceprints.

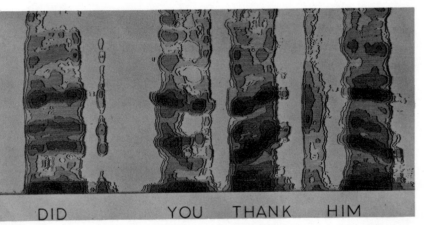

Fig. 7.9. Voiceprint using the contour method of indicating intensity. Courtesy Kay Electric Co.

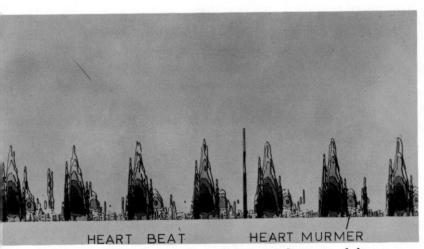

Fig. 7.10. Print of heartbeat showing indication of heart murmer. Courtesy Kay Electric Co.

8.

Audiometry and Noise Hazard

8.1. *Introduction.*

In primitive civilizations noise was probably not a factor in damage to hearing but it probably was, at times, an annoyance factor and a factor interfering with conversation. With the birth of the Industrial Revolution, however, factories with high noise levels were developed and there were many instances where workmen were exposed to noise levels which caused hearing damage. With the advent of jet aircraft and rock 'n' roll groups using high-power amplifiers, we now have new noise hazards not necessarily connected with an individual's employment.

The threshold of hearing varies with the frequency of the sound. Figure 8.1 shows a plot of the normal threshold of hearing as a function of frequency. This is an idealized plot made by averaging the results measured on a number of subjects. The threshold plot made on a single subject would not yield such a smooth curve. A plot such as that shown in Fig. 8.1 is called an audiogram. In order to obtain an audiogram for a subject, special precision equipment must be used and the measurements must be made under conditions where there is no extraneous noise to interfere with the tests. The tests are made using pure tones of the various frequencies over the audio or sonic range.

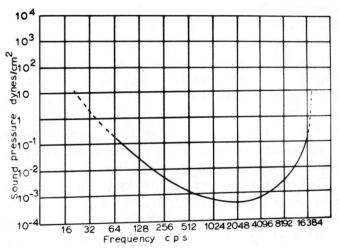

Fig. 8.1. Threshold sound pressure as a function of frequency for average normal human ears.

8.2. *Masking.*

If we are listening for a weak sound in the presence of another sound, the hearing threshold for the desired sound will be raised due to the masking effect of the extraneous sound. Masking sounds may be pure tones or noises in broad bands. They may be continuous or intermittent. A resident near an airport is subject to the masking noise of aircraft taking off and landing. This is an intermittent broad band noise which, in some instances, may be so loud that all conversation must cease during the take offs and landings of the aircraft. A similar situation may exist in homes near a heavily traveled highway.

An interesting example of masking is the so called

"party effect." When a large number of people are assembled in a room and all attempt to converse at the same time, the masking effect of the voices causes each individual to raise his voice until the condition is reached where everyone is speaking as loud as possible.

If a masking sound occurs in a very narrow frequency range it will have a maximum masking effect for sounds in that general range of frequencies. A listener can normally distinguish a wanted sound in the presence of masking sounds of higher level. To do this he utilizes frequency differences, the directional characteristics of his ears and the signal processing system in his brain.

We generally consider masking noise as a nuisance, but it can be a hazard even though the level is below that which would cause damage to the hearing mechanism. If the masking noise prevents a person from hearing a danger signal or instructions for avoiding danger, it is a hazard. Such noises can cause frustration, contribute to fatigue and even upset the psychological well being of people.

8.3.   *Hearing Threshold Shifts Due to Noise.*

Figure 8.1 shows the average normal hearing threshold as a function of frequency, as measured in the absence of masking noise. If the subject is exposed to an excessive noise level for a period of time and an audiogram is taken immediately afterward, the threshold will be found to shift upward by several dB. The first question which occurs to us is, What is an excessive noise? This seems to be a factor that varies with the individual, the duration of the exposure and how continuous the exposure is. An excessive sound level may be as low as 85 dB as read on a sound level meter or it may be higher than 100 dB. The hearing threshold will usually return to normal after a rest period, and in such instances we designate the shift as a *temporary threshold shift.* If the hearing threshold does not return to normal we say that the subject has suffered a *noise induced permanent threshold shift.*

There are many factors that make it difficult to obtain accurate information about the degree of hazard involved in a particular noise environment. Every individual suffers some loss in hearing acuity with age, particularly in the high-frequency region, even though he does not suffer exposure to high noise levels. When people work in a factory under high-noise-level conditions, one might expect that it would be possible to obtain accurate information on the effect of the factory noise on their hearing sensitivity by periodically taking their audiograms. However, there is usually no means of controlling the noise environment to which they are exposed outside of working hours. They may spend considerable time dancing to rock 'n' roll music, where the noise levels are 20 to 30 dB higher than those in their working environment. They may travel to and from work by a means of transportation that generates excessive noise levels, and there is also the problem of individual differences in susceptibility to noise.

On the basis of tests which have been carried out, the amount of temporary threshold shift in dB is proportional to the logarithm of the time of exposure to the noise. This means that each doubling of the exposure time will result in the same number of dB shift in the threshold. The recovery from the threshold shift in the absence of the noise follows the same law and recovery will be complete in most instances within 16 hours after the subject is removed from the noise environment. However, If the threshold shift is between 40 and 50 dB it may require several weeks for complete recovery.

Intermittent noise is much less serious in producing temporary threshold shift than continuous noise. For example, if a given noise level is present only half of the time during a given work period, the temporary threshold shift will be less than half as great as it would be if the noise is continuous. It has also been shown that when people work in a continuous noise environment the amount of temporary threshold shift is considerably reduced when the subjects take one or two 15-minute breaks away from the noise during each half day.

Low-frequency noise is less likely to produce threshold shift than high-frequency noise of the same level. A rumble is much less damaging than a squeal at the same level. If the offending noise occurs in a narrow band of frequency, the maximum threshold shift will occur at a higher frequency, usually from one-half to one octave higher than the offending frequency. The hearing threshold in the high-frequency regions shifts with age but the aging effect varies greatly with different individuals. If the principal noise in the working environment is in the neighborhood of 4000 cycles per second it is sometimes difficult to be certain that a permanent loss of high-frequency sensitivity is due to the noise exposure or to the aging effect for the particular subject.

No known means of medical treatment will affect the amount of threshold shift a given individual will experience when he is exposed to noise. If a permanent threshold shift is produced there is no known method of treatment to restore it. The only known factors that can control the amount of temporary or permanent threshold shift are the level of noise, the duration of exposure and the susceptibility of the individual to the noise. Some observers believe that individuals who show a high sensitivity to temporary threshold shift will have a correspondingly high sensitivity to permanent threshold shift. This is not necessarily true. Although there probably is a relation between the two kinds of shifts, the relationship may not be so close as originally assumed.

In the early period of development of factory industry, little or no attention was given to the effect of the noise environment on hearing. Apparently both the employer and the employees were willing to accept the fact that the employees in some factories would become completely deaf after a few years of working in the factory. It is only relatively recently that concern has developed for the preservation of the hearing of industrial workers. This is partly due to a more enlightened social attitude and the fact that hearing loss on the job is now recognized by the courts as grounds for payment of workman's compensation. With the improvements in machine technology, it is mechanically possible to do machine operations at higher speeds than were possible in the past. These higher machine speeds tend to result in higher noise levels at higher frequencies.

8.4. *Acceptable Noise Exposure.*

A noise exposure which produces an appreciable permanent shift in the hearing threshold is not acceptable. A permanent shift, however, may not become evident until the subject has been exposed to the noise environment for several years. The noise exposure depends on the noise level, the frequencies present in the noise, the intermittancy of the noise, the number and duration of rest periods, the length of the working day and the years of exposure.

It is possible to define hearing impairment as a condition such that it is difficult to hear everyday speech. This is difficult to measure but has been defined in terms of the hearing threshold for pure tones at 500, 1000 and 2000 cycles per second. If the threshold for all three of these frequencies is not higher than 25 dB on the scale shown in Fig. 2.3, it is satisfactory to assume that the subject has not suffered hearing impairment, because most normal subjects will have thresholds between 0 and 25 dB on this scale. If the subject's thresholds for these frequencies are higher than 40 dB on this scale he will require amplification to readily hear everyday speech.

Since permanent threshold shift is an irreparable damage to the hearing mechanism, we should not wait until it has occurred to decide that the noise level is unacceptable. The problem is further complicated by the fact that there is considerable difference in the amount of noise exposure which will produce hearing damage in different individuals and there is a great difference in the amount of noise to which different individuals are exposed when they are away from work. The worker may be a member of a rock 'n' roll group or he may be a member of a shooting club and do skeet shooting on week ends. These outside activities may provide more damaging noise exposure than that which he encounters during his working hours. The legal determination for compensation places the burden of liability on the employer when the employee works in a high noise environment and no allowance is normally made for the outside activities of the employee. It is therefore desirable to have some means of determining when a risk to the employees hearing is involved before the onset of permanent damage.

It has been suggested, on the basis of experience, that the amount of temporary threshold shift, measured two minutes after an eight-hour exposure to the noise in the working environment, is an indication of the amount of permanent threshold shift that would be produced after 10 to 27 years of daily exposure to the same noise environment. If the temporary threshold shift measured at 2000 cycles per second under these conditions is not more than 10 to 12 dB, the noise environment may be considered to be reasonably safe. However, tests should be made periodically on workers exposed to such an environment to determine if they are experiencing any permanent shift in their hearing threshold. It is important that these tests be made after a sufficient rest period away from the noise to enable recovery from any temporary threshold shift. If the temporary shift in threshold produced in one working day has not completely recovered by the time the employee reports for work the next day, the worker is probably receiving excessive noise exposure.

There is apparently more individual variation in susceptibility to damage from impulse noises than from continuous noise. The old-time boiler factory worker and blacksmith were exposed predominantly to impulse noises. People who do considerable skeet shooting are, of course, exposing themselves to severe impulse noises and there is a large amount of impulse noise involved in the modern dance orchestras.

8.5.   *Industrial Noise Control.*

Since there is now considerable knowledge about the hazard due to noise, and there is also the recognition of the employer's legal responsibility to compensate employees for hearing damage sustained in their working environment, considerable attention is being given to noise reduction in industrial machinery. This effort is hampered by the lack of engineers with adequate knowledge of acoustics. The mechanical engineer who designs a machine is apt to think in terms of the mechanical process that the machine is to perform and he will design the machine to perform that process, but his lack of knowledge of acoustic principles can result in a machine that will generate higher noise levels than it would if adequate attention had been given to the noise problem during the design process. It is very difficult to reduce the noise of the machine after it is designed and constructed. Unfortunately there are not enough acousticians available to assist the engineers who design the machines. There are instances where industrial processes must be operated at a slower rate than would otherwise be possible in order to maintain an acceptable noise environment for the people who operate the machines.

In addition to the hearing hazard, high noise levels in the factory environment interfere with speech communication. This interference may cause loss of efficiency and can even result in other physical hazards because the workers may fail to hear danger signals.

8.6.   *Personal Protection.*

If the sound level in a work area exceeds safe values the personnel in the area can obtain protection by means of ear protectors. Such ear protectors may be ear plugs prefabricated of rubber or plastic, earmuffs designed to completely cover the external ears or helmets designed to cover the entire head of the worker. Ear plugs are made in a variety of sizes and must be carefully fitted to each individual. An improperly fitted ear plug is practically worthless for ear protection. Ear muffs are rigid cups with a cushion filled with plastic foam or a fluid to seal against the skin surrounding the external ear and held in place by means of an adjustable head band. Ear muffs are often equipped with transducers so that they can serve as headphones as well as ear protectors. Helmets are often used when the worker needs protection against bumps or flying missiles in addition to sound. They are usually equipped with transducers at the ears and a microphone to facilitate communication.

When properly fitted ear plugs or ear muffs are used, sound conduction by the normal path via the middle ear is virtually eliminated but sound conduction by the bones of the head is still permitted. For this reason, the reduction of sound level at the inner ear is limited to about 60 dB at high frequencies and about 35 dB for frequencies below about 250 cycles per second. A helmet can produce about 10 dB more reduction after which conduction of sound by the whole body limits the possible attenuation.

Ear plugs, properly fitted, can reduce sound levels to safe values in sound fields up to 110 dB as measured on the A scale of a sound level meter and ear muffs can be used in sound fields up to 125 dB. The sound levels at which these protectors can be safely used, however, can be increased if the time of exposure is reduced.

All ear protectors are uncomfortable. Usually ear muffs are less uncomfortable than ear plugs, although they are heavier and more bulky. In areas where ear protection is necessary, close supervision is often required to be certain that the workers wear their ear protectors. The discomfort of the ear protector often makes the worker prefer to risk his hearing rather than wear it. This is not an uncommon human reaction, as evidenced by the fact that workers will often avoid wearing safety glasses in places where there is an eye hazard.

The method most commonly used to evaluate ear protectors is to measure the shift in the hearing threshold of several individuals when the ear protector is applied. This assumes that the sound attenuation for high level sounds is the same as it is for the low-level sounds that must be used for such a test.

The use of ear plugs or ear muffs does not preclude voice communication. If the noise level in the environment is relatively low, hearing of voice communication at three feet will not be seriously degraded (See Fig. 2.3.). When the noise level is high, the masking effect of the noise on voice communication will be no worse with ear protectors than without them, and in some instances the masking effect will be reduced when ear protectors are worn.

8.7. *Legal Aspects of Noise Hazard.*

It is now generally recognized that hearing loss is a disability for which an employee can claim compensation. Unfortunately for employers, there are many instances where employees can claim compensation for hearing disability that was not incurred at the place of employment. In order to protect themselves, employers need to institute hearing test programs in order to provide the complete history of any variation in hearing thresholds of their employees from the time of their initial employment.

The Federal Government and several of the states have enacted legislation attempting to define conditions of noise hazard and methods of measurement. These laws must be administered by the public health services. The problem of administering these laws is made difficult because of the shortage of trained acousticians.

8.8. Community Noise.

Community noise usually does not constitute a hearing hazard but is often a nuisance, contributing to frustration and sometimes psychological upsets of the residents. Some of the countries in Western Europe recognized this condition much earlier than we have in the United States. For example, a manufacturer who intends to market any type of motor vehicle in West Germany must subject the vehicle to government tests and he is not permitted to market the vehicle until it meets the noise standards specified by the government.

Community noise is usually caused by traffic, aircraft, nearby factories, music halls and even radios and television sets operated at high volume. With the advent of supersonic aircraft, booms are produced which are disconcerting and sometimes cause damage to structures and may even cause hearing damage.

The most common source of community noise is caused by traffic. The noise levels will be higher if there is a hill than if the roadway is level. Trucks going up the hill require the application of high power causing a high noise level while trucks going down the hill are subject to backfiring. I have stayed at a hotel located on a highway passing through the city where the hotel management had to supply ear plugs for the guests in the rooms at the front of the hotel to enable them to sleep.

Considerable study has been made in West Germany of methods of minimizing traffic noise in towns and cities. They have found that high hedges planted next to the street are quite effective in reducing the noise transmitted into the houses.

The problem of aircraft noise is of much importance in a city like New York. In order to minimize the effects of aircraft noise the aircraft that land and take off from some of the airports must meet certain noise standards and often there are limits imposed on the amount of thrust that can be generated until the aircraft has passed a certain distance beyond the city's populated area.

Noise is now classified as a form of pollution and many studies are in progress to establish reasonable limits on community noise and to develop methods of noise reduction and control. Unfortunately noise—like the other kinds of pollution—tends to increase with technological advances and population increase. Considerable effort is necessary just to keep the noise levels in our communities from increasing. It may be necessary to incorporate noise isolation in our houses and apartments in order to secure the desired quiet.

9.

Sound Recording and Reproduction

9.1. Early Methods of Sound Recording and Reproduction.

Sound recording consists of using the sound pressure or sound velocity variation to produce a mechanical or other physical variation in the recording medium, which can later be converted to sound waves. At the time of the invention of sound recording, electronic amplifiers were not available so the process had to be an entirely mechanical one. Figure 9.1 shows the method of sound recording Edison used. Figure 9.1 A shows the elements of the sound recording system.

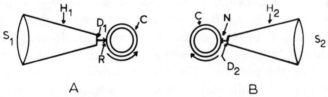

A   B

Fig. 9.1.   Original Edison sound recording and reproducing system. In A, S_1 is the input sound, H_1 is the recorder horn, D_1 is the recorder diaphragm, R is the recording cutter and C is the cylinder on which the recording is cut. In B, C is the cylinder on which the recording was cut, N is the reproducing needle, D_2 is a diaphragm, H_2 is a horn and S_2 is the sound output.

The source of the sound was positioned at the mouth of the horn and the small end of the horn was closed with the diaphragm D_1. The cutter R was designed to cut a groove in the wax cylinder C as it rotated and the cutter was advanced parallel to the axis of the cylinder to cut a spiral groove on the surface of the cylinder. The varying sound pressure on the diaphragm caused the cutter to move so that it would cut a variable depth groove with the depth variations proportional to the sound pressure variations and the frequency of the hills and dales on the bottom of the groove corresponded to the frequencies present in the sound signal. Since the area of the mouth of the horn was large compared with the area of the end carrying the diaphragm, it served to match the impedance of the air outside the horn with the higher impedance of the cutter operating on the wax cylinder. The pressure variations on the diaphragm were, therefore, higher than the pressure variations at the mouth of the horn and the particle displacements were correspondingly less.

When the cylinder was placed on the reproducer, shown in Fig. 9.1 B, a needle was forced to follow the groove cut by the cutter by being driven along the cylinder by means of a screw of the same pitch as that used on the recorder. The needle moved back and forth, following the hills and dales on the bottom of the groove, actuating the diaphragm D_2 at the frequencies of the hills and dales in the groove. The horn H_2 served as an impedance matching device to match the high impedance of the needle and diaphragm to the low impedance of the surrounding air.

This was an extremely simple system, but it worked. The range of frequencies the system could handle was limited but it could handle the range of voice frequencies reasonably well. The wax used in making the original impression was relatively soft, so the record would deteriorate rapidly when played. Commercial records were made by using the wax recording as a master. This master was used to produce a metal negative from which impressions could be made using a harder material. The material used at that time was a shellac-base plastic that could be pressed when hot and was sufficiently hard so that a record could be played back hundreds of times without serious deterioration. The original Edison phonographs used a diamond-tipped needle insuring a long life for the playback needle.

One serious disadvantage of the cylinder recordings was the storage space they required. This led to the development of the present-day disc record. At the same time, the nature of the recording groove was changed so that, instead of the recording consisting of hills and dales on the bottom of the groove, the cutter was vibrated perpendicular to the groove and parallel to the surface of the disc to produce a wiggly groove of

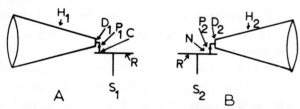

Fig. 9.2.   Arrangement for making lateral disc recordings. In the recorder at A, D_1 is the diaphragm, P is a pivot, C is the cutter and the disc R is rotated by the shaft S_1. In the reproducer at B, the record R is rotated by the shaft S_2, N is the needle, P is a pivot and D_2 is a diaphragm.

constant depth. Figure 9.2 A shows the arrangement for making a wiggly groove or lateral recording. The recording is made by cutting a helical groove of constant depth in a wax covered disc. A screw is used to advance the cutter along the radius of the disc and the vibrations of the diaphragm D_1 cause the cutter to vibrate perpendicular to the groove at the frequencies of the sound and at amplitudes corresponding to the sound pressure.

The lateral recorded disc playback system is shown in Fig. 9.2 B. The needle N follows the groove and the wiggles in the groove cause the needle to vibrate the diaphragm D_2, generating the sound which is radiated from the mouth of the horn H_2. A screw is usually not used to advance the needle across the record, as the groove itself can be used to track the needle. For many years needles of soft steel, wood and even cactus thorns were used, and one needle could be used for playing only one or two records.

9.2. *Modern Phonograph Disc Recording and Reproduction.*

With the invention of the vacuum tube and the development of electronic amplifiers, the way was paved for great improvements in the recording and reproduction of sound. In the modern recording process the recording cutter is electrically driven from a suitable power amplifier. The sound signal is picked up by a microphone in a convenient location and transmitted by cable to the microphone amplifier, which amplifies the signal and feeds it to the power amplifier, which drives the cutter. Any defects in the frequency response of the system can be compensated for by means of filters in the amplifiers.

The modern master records are metal discs with a thin layer of a special plastic that will leave a smooth surface when cut by the cutter. The negative is made of metal and two negatives are used to press recordings on the two sides of the final disc. Originally a shellac base material was used for the commercial discs but the modern recordings are pressed from a plastic material that produces a much smoother surface in the groove. Any roughness in the groove is reproduced as noise in the reproducer so a smooth surface on the walls of the groove is important.

The use of electronic amplifiers in the sound reproducing system has also led to important advances. Instead of depending on a needle following the groove in the record to drive a diaphragm, we have available several transducers operating on different principles, all of which are able to convert the motion impressed on a stylus following the groove into electrical signals that can be amplified and converted to sound by means of a loudspeaker.

In order to reproduce high frequencies without having to exert excessive forces on the sides of the record groove, it is necessary that the amplitudes of the groove wiggles be as small as possible and the mass of the stylus and its attachment to the transducer be small. As the force that must be exerted on the stylus by the sides of the groove is increased, the tracking pressure on the stylus must be increased in order to prevent it from jumping out of the groove.

It is important to stop and calculate the magnitude of the pressure exerted by the point of the stylus in the record groove. The shape of the stylus is shown in Fig. 9.3 where the end of the stylus which rides in the groove

Fig. 9.3. Shape of a modern phonograph stylus.

is a hemisphere of a radius from 0.0005 to 0.001 inch, with 0.00075 inch the most common. The tracking force used in modern phonographs will vary from 1 gram (0.035 oz.) to 5 grams (0.175 oz.). Let us assume that a tracking force of 5 grams is used with a stylus of 0.00075-inch radius. The area of the hemispherical end of the stylus is $2\pi \times (0.00075)^2 = 7 \times 10^{-6}$ square in. (7 millionths of a square in.). The pressure, which is the force per unit area, is equal to $0.175/(7 \times 10^{-6}) =$ 25,000 oz. per square in. or 1600 lbs. per square inch. This is the static pressure in a groove with no recorded signal. When there is a signal recorded in the groove the pressures will be larger depending on the frequency and amplitude of the signal. This illustrates the importance of developing transducer stylus combinations of small mass, which require small displacements to generate the required signal. A further requirement is that the transducer-stylus combination have no resonant frequency within the sonic range.

There are two general categories of phonograph transducers: magnetic and electrostatic. The magnetic transducers (usually called pickups) may involve moving a coil of wire in a magnetic field produced by a fixed magnet or a means of shifting the magnetic flux from one coil to another. The latter type is illustrated in Fig. 6.8.

The electrostatic pickups usually use ceramic elements

or piezoelectric elements to generate a voltage proportional to the deflection of the stylus.

The voltage generated in a magnetic type of pickup is proportional to the rate of change of flux in the coil and will therefore be proportional to the side to side velocity of the stylus. This means that the voltage generated by a given amplitude of wiggle of the stylus will decrease as the frequency decreases. In order to correct for this the phonograph amplifiers used with magnetic pickups have electrical networks incorporated which provide a gain increase of 6 dB per octave from midrange to the low-frequency end of the sonic region.

In order to minimize the stylus forces at high frequencies, the high-frequency region is recorded at amplitudes on the record which decrease as the frequency increases above the midrange and compensating networks are incorporated in the amplifier to correct for this condition.

No low-frequency correction is needed with electrostatic pickups and the signal voltage generated in them is higher than those generated in magnetic pickups. The networks for low-frequency compensation are located in the first stages of the amplifier and two sets of terminals are provided—one for the magnetic pickups and a second, which connects to the amplifier circuit beyond the low-frequency compensating circuits, for the electrostatic pickups. In addition, the more sophisticated amplifiers have networks that can be controlled externally for varying the relative gain of the amplifier at the low and high frequencies to compensate for the frequency response of the speakers and the acoustics of the room. Some amplifiers even have a "loudness" control, which corrects for the fact that at low levels the response of the human ear is less at low frequencies than it is in the midrange of frequencies. This control makes it possible to play records at very low levels without the apparent loss of low-frequency response caused by the characteristics of the ear. It is obvious that when a magnetic pickup and loudness control are used it is difficult to achieve adequate shielding to prevent 60 cycle pickup from the motor that drives the record and rumble caused by irregularities in the drive of the record.

The availability of electronic amplification has provided many advantages to make possible high fidelity reproduction of music. It also provides a disadvantage since a high-power amplifier makes it possible to generate high levels of sound, which can contribute to the nuisance of community noise.

One of the features of the ordinary phonograph record that contributes to unreality is that music from a large group—such as a symphony orchestra—is heard over a large angular range when the orchestra is listened to at a live concert but is heard from a single point when the reproduction is played on a phonograph. It was recently demonstrated that if two recordings are made using microphones spaced near the two sides of the ensemble, and these recordings are played back simultaneously through two amplifier and speaker systems with the speakers spaced so that their angular separation for the listener corresponds to the angular separation which the listener would experience when listening to the performance live, the reproduction would give the illusion of a stereophonic effect and the reproduction would sound much more realistic.

The first stereophonic recordings were made by using

two sets of grooves with two tone arms locked together. Later it was found that the two signals could be incorporated in a single groove with one signal effectively recorded as a hill and dale recording and the other as a lateral recording. A single transducer can be designed to pick up and keep separate these two signals and they are fed to two separate amplifiers and speakers. Stereo reproduction systems are becoming so common now that many of the record companies manufacture only stereophonic records. These records can be played on a monophonic system, although the latter will respond only to the lateral component of the recording.

One of the errors which leads to distortion of the recorded signal is caused by the fact that the stylus must have a finite radius. At high frequencies, such a stylus will trace a path which is not an accurate trace of the wiggles in the groove. In order to correct for this effect, a system has been devised which distorts the signal being recorded in the inverse manner of the distortion produced by the stylus. By this means the stylus tracing distortion is minimized.

9.3. *Optical Recording and Reproduction.*

Talking motion pictures were made possible by the invention of optical recording. This enabled the soundtrack to be recorded directly on the film. Attempts to have the recordings made separately on discs were not practical because it was too difficult to keep the sound synchronized with the picture in the theatre.

The modern-day soundtrack is recorded on a narrow strip at the edge of the film, by using either of two methods. One is called the *variable area* and the other the *variable density* method of recording. In the variable area recording, an image of a triangular light source is formed on a slit after reflection from a galvanometer

Fig. 9.4.   Variable area sound recording system. The illu-
minated area L is focused on the slit S after being
reflected from the galvanometer mirror.

mirror as in Fig. 9.4. The amplified signal from the microphone forces the galvanometer mirror to oscillate, moving the triangular bright area up and down over the slit and thereby varying the length of the slit, which passes light. A lens forms an image of the slit on the film. This forms a black image on the negative whose width varies at the frequency of the signal and the variation of amplitude of the width is proportional to the intensity of the signal. On the positive film this black image becomes transparent. The sound is reproduced by placing a light source on one side of the film, and the light transmitted through the film in the area of the soundtrack falls on a photoelectric cell whose output voltage is amplified and fed to the speakers. The stereophonic effect can be produced by using a double soundtrack and using two sets of amplifiers and speakers.

In the variable density system, the light is modulated by a light valve consisting of a pair of ribbons in the light path. This system produces a track of fixed width and variable density. The negative film will have minimum transmission at maximum level but the positive print will have maximum transmission at maximum sound level.

Due to the high rate of film transport necessary for motion picture reproduction, it is possible to achieve high fidelity of sound reproduction of optical recordings. The high cost of film and film processing makes optical recording impractical for other than motion picture sound but it is a very convenient method of reproduction of sound with motion pictures as a very narrow portion of the film is sacrificed to provide the soundtrack and the soundtrack becomes an integral part of the film. Since the film is jerked through the projection gate one picture at a time, the sound image synchronized with the picture is displaced so the sound reproducing system views a region on the film where it is being driven at a uniform speed.

9.4. *Magnetic Recording.*

The first magnetic recording was done on steel wire.

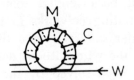

Fig. 9.5.   Wire recorder transducer.

Figure 9.5 shows the type of transducer used to record the magnetic image on the wire. The wire W is passed over the gap between the polepieces of the magnetic core M which has a coil C wound toroidally on it. The coil carries an electric current which is the amplified signal from a microphone. Therefore, the magnetic flux at the air gap in M will vary with the frequency and amplitude of the signal. For a given wire speed and air gap width there will be an upper frequency at which the amount of magnetization of the wire will decrease. This frequency limit can be raised by either increasing the speed of the wire or decreasing the width of the air gap. The signal can be played back by passing the wire over a similar transducer so that its magnetic signal will vary the flux in the air gap, causing a voltage to be generated in the coil. This voltage can be amplified and fed to a speaker.

Since the voltage generated by a varying magnetic field is proportional to the rate of variation of the field, there will be a falloff in playback response at low frequency similar to that in a magnetic phonograph pickup. Compensating networks will therefore be necessary to bring the low-frequency response up to that in the midrange of frequency. For a given wire speed and gap width there will be some upper frequency at which the response will begin to fall off in level. The upper frequency range can be extended somewhat by use of frequency compensating networks in the amplifier.

The wire recorder suffers from a disadvantage: it is difficult to handle the wire without danger of breakage. This problem has been solved by the use of magnetic tape, which is prepared by depositing a thin layer of magnetic material on a thin plastic tape. The magnetic tape is easy to handle and it has a finite width so that more than one channel of magnetic signal can be recorded on the same tape. This provides an easy means of recording the two signals for stereophonic playback.

Magnetic tape is available in widths from ¼ inch to 2 inches and from one to fourteen channels can be recorded in parallel simultaneously.

The recording and playback transducers used with magnetic tape are similar to that shown in Fig. 9.5. Although the audio signal can be recorded as described above, it is necessary to apply a high-frequency biasing signal along with the audio signal being recorded in order to obtain acceptable recordings. The biasing signal is at a frequency considerably higher than the range of frequencies being recorded and it is mixed with the recording signal in the same coil on the transducer.

The tape speeds used in recording sound signals vary from 1.7 inches per second to 15 inches per second, depending on the frequency range desired. For intelligible speech recording and reproduction a speed of 1.7 inches per second is adequate. For reasonably high fidelity reproduction of music 7.5 inches per second is usually used and for the best high-fidelity reproduction a speed of 15 inches per second is used.

The use of magnetic tape recording has been an important asset to the recording industry. It is relatively simple to make high-fidelity recordings on magnetic tape, which also has the advantage of being relatively easy to edit. If someone plays off tune in a recording it is not necessary to rerecord the entire performance. The bad portion can be rerecorded and a new section of tape spliced into the original tape. When a satisfactory tape recording is available, it can be used as the source of signal for making the master disc recording. This recording can be done in a laboratory under carefully controlled conditions and, as the tape can be played over a large number of times without deterioration, a number of trials can be made if necessary to obtain a satisfactory master record. Some recording engineers produce the master record from tape recordings which were made at a speed of 15 inches per second by operating the tape playback and the master record at half speed, thereby reducing the frequencies to which the cutter of the master record must respond by a factor of two. By this procedure, if the maximum frequency to be recorded is 16 kc the maximum frequency at which the cutter must operate will be reduced to 8 kc. The records made from the master will, of course, be operated at their standard speed, which will be twice that at which the master was operated during the cutting operation.

This versatility of magnetic tape recording is utilized in sound analysis. For example, there are instances where analyses of subsonic frequencies are to be carried out. Frequencies one or two octaves above one cycle per second can be recorded on magnetic tape. If these signals are recorded on tape running at 1.87 inches per second and played back at 15 inches per second, the frequencies recorded will be increased by a factor of eight so that analysis can be more readily carried out.

Since the magnetic tape recording is a magnetic recording of an electrical signal generated in an electroacoustic transducer, it is possible to use the magnetic tape to record any other kind of electrical signal. Magnetic tape recorders can therefore record the video signals of television. Many of our television programs are, of course, prerecorded on magnetic tape, which has made possible the "instant replay" in televised football games.

It is necessary that the signal supplied to the magnetic

tape transducer be AC. However, DC signals or very low frequency AC signals can be recorded by allowing them to modulate an AC signal generated by a fixed frequency oscillator. The signal from the reproducing unit can then be demodulated to recover the original signal.

Nearly all sound analysis is done by prerecording the signal to be analyzed with the necessary calibration signals, after which the laboratory analyses are carried out. A high-quality tape recorder is much more easily portable than the analysis equipment.

Magnetic tape recorder-reproducers for home use can be purchased for less than $100, but high-quality instruments capable of use at high frequencies for sound analysis may cost from $5000 to $20,000. Some of these units may record as many as 14 channels on a wide tape so that signals from several sources can be recorded simultaneously, leaving a channel for the engineer to record voice comments about the conditions under which the data were recorded.

9.5. *Distortion.*

Any difference between the reproduced sound and the sound from which the recording was made is called distortion. The aim in high-fidelity music reproduction is to attain a reproduced signal as identical as possible to the original signal. The distortion may occur either in the recording phase or in the reproducing phase.

In the early phonographs, frequency distortion was the most obvious because the mechanical system was incapable of recording and reproducing the entire sonic range of frequencies. Many systems in use today have considerable frequency distortion. For example, ordinary AM radios in use today limit the highest transmission frequency to 5000 cycles per second. Many of the loudspeakers in use are not capable of reproducing frequencies lower than 75 to 100 cycles per second. In fact, when many of the small speakers are excited with frequencies below about 75 cycles per second they will respond with the second harmonic of the exciting frequency.

When the first attempts were made to achieve high-fidelity reproduction of music, the attention of the engineers was focused most strongly on the frequency distortion and they were surprised when experienced listeners preferred to listen to reproduction which limited the frequency range. The reason for the preference for reproduction of a narrower band of frequency was that in achieving the higher fidelity of frequency range, other distortions were increased or became more obvious, and these other distortions were more displeasing to discriminating ears than the absence of some frequency range at the high and low ends of the spectrum.

The most obvious of these other distortions is harmonic distortion. When harmonic distortion is present, an original pure tone, when reproduced, will in addition to the original frequency contain frequencies that are integral multiples of the original. If the single frequency of 500 cycles per second is recorded, the reproduced signal may in addition to this frequency contain those of 1000, 1500, 2000 and 2500 cycles per second. To the non-musical engineer the effect of this harmonic distortion is not disturbing because these harmonics are not discordant. However, the unique characteristics of the sounds produced by the various musical instruments are determined by the particular harmonics which they pro-

duce and the relative intensities of these harmonics relative to the fundamental frequency. If middle c is
sounded on a clarinet and then on a violin, the frequency
of the fundamental tone will be the same in both instances, but little experience is required for a listener
to distinguish between the sounds of the two instruments. If the reproducing system contributes considerable harmonic distortion, the reproduced signal originally produced by a violin may sound the same as that
originally produced by the clarinet. This may not be
disturbing to the inexperienced listener but it is quite
disturbing to the experienced listener who wants each
instrument to sound the same in the reproduction as it
does live.

Even if the signal recorded on the disc record or on
a magnetic tape is free from distortion, harmonic distortion can occur in the transducer which detects the signal, or in the amplifier or in the loudspeaker. None of
these elements is ever completely free from harmonic
distortion but engineers are continually striving to minimize it. It is more difficult to achieve a given reduction
of harmonic distortion in a system that reproduces a
nine-octave band than in one that reproduces a six-
octave band. In order to minimize the distortion, very
high quality must be maintained in all elements of the
system. This is one reason for the high cost of truly
high-fidelity-reproducing systems.

Another type of distortion is called *intermodulation*
distortion. The name of this distortion sounds complicated but it is an accurate description of the process
which causes it. All of the elements in the system must
reproduce a large number of frequencies simultaneously
and it is difficult to design these elements so that their

sensitivity is the same over their entire range of operation. If their sensitivity is not the same over the entire range they are said to be non-linear. Any portion of the system can be non-linear but we can visualize what happens more easily if we consider the speaker. In Fig. 6.1, the coil in the speaker will be deflected in the magnetic field by the various currents corresponding to the various signals being reproduced. Suppose that the speaker is reproducing two frequencies, one of 50 cycles per second and the other of 5000 cycles. During one half-cycle of the 50-cycle-per-second signal there will be 50 cycles of the 5000-cycle-per-second signal, but if the speaker is non-linear in its response the amount of deflection produced by the 5000-cycle-per-second signal at the time the coil reaches its maximum deflection due to the 50-cycle-per-second signal will be less than the corresponding deflection at the time when the current for the 50-cycle-per-second signal passes through zero. The 5000-cycle-per-second signal is therefore *modulated* by the 50-cycle-per-second signal.

When one signal is modulated by another, two additional signals are produced. One of these has a frequency equal to the sum of the two and the other is equal to their difference. In this instance we will have frequencies generated at 5050 and 4950 cycles per second. If we are reproducing the sound from a symphony orchestra, in which a large number of frequencies must be reproduced simultaneously, we can have many of these sum-and-difference frequencies, which are not harmonious with the original frequencies or with each other.

Harmonic distortion is objectionable because the reproduced sounds do not resemble the sounds of the live instruments, but the additional frequencies generated are harmonious. The intermodulation distortion is more objectionable because the additional frequencies generated are not harmonious. To reduce intermodulation distortion to a minimum, all elements of the system must be made to have a linear response over the entire range of operation.

Sudden bursts of sound are called transients. The transients from percussion instruments and a piano are quite severe—the sound intensity builds up from zero to its highest level in a very short time. If percussion instruments are to sound natural, it is necessary for the sound power output of the speaker to build up to the full level in the same time as the original sound was built up to its full power, and the system must do this without overshooting. We may have a system that shows very low distortion when tested with steady tones but has a poor transient response. When listening to a demonstration of a phonograph system, a good kind of test record is a piano concerto. Listen to the recording and note if the piano sounds like a live one.

Distortions are particularly important when recordings are made for sound analysis. Harmonic and intermodulation distortion can cause frequencies to appear in the analysis which were not present in the original sound. For this reason, the tape recorders used for recording sounds for later analysis must be of the highest quality and must be maintained in peak operating condition. To minimize errors due to distortion, use is made of multi-channel recordings so that the number of octaves of frequency recorded on each channel is not large enough to produce unacceptable distortion.

10.

The Physics of Music

10.1. *Introduction.*

Music was created and enjoyed long before there was any knowledge of the physical principles involved. We can give a physical description of a musical sound in terms of the frequencies present, their relative intensities, the absolute level of the sound and its variation in level with time. A listener, however, considers it a musical sound because of his subjective reaction to it. The knowledge of the physical description of the sound will not make the listener any more appreciative of it.

We are interested in the physics of music because it can help us to understand the operations of musical instruments, and with a knowledge of their operation we can learn how to construct better instruments and we can also learn how to adapt new materials in the construction of musical instruments.

There is a tendency for acousticians to feel that musicians know nothing about acoustics. The musician, however, knows a great deal about acoustics but his knowledge is expressed in terminology so different from that of the physicist that it is virtually impossible for them to communicate. The physicist thinks of sounds in terms of the measurements he can make with his physical instruments while the musician depends on the finest of all sound measuring instruments, the trained human ear, to evaluate musical sounds. The ear does not measure the characteristics of sounds on the same scales as those used by the physicist. The scale of pitch is not the same as the scale of frequency. The scale of loudness is not the same as the scale of intensity level, and while the scale of intensity level is the same at all frequencies, the scale of loudness varies with frequency.

Young people now are not willing to serve years of apprenticeship to learn the skills necessary to build fine musical instruments as they did in the time of Stradivarius. In order to train craftsmen to build such instruments, we need to be able to describe accurately, in terms of physical measurements, the difference between a good and a poor instrument. We can then prescribe physical tests to be applied during its construction to replace the tests by ear used by the early craftsmen.

10.2. *Pitch.*

The frequency of a sound is an objective characteristic, the number of cycles per second. We can measure the frequency with physical instruments. Pitch is a subjective characteristic, one we can sense with our ears. If we generate a pure tone of a given frequency and then increase its frequency, we will sense an increase of the pitch. However, if an observer is asked to select a pitch half that of another pitch, the frequency he selects will not be half the frequency corresponding to the original pitch. There will, of course, be some disagreement between different observers but a scale of pitch constructed in this manner will not have a linear relationship with the physical scale of frequency.

The unit of subjective pitch is called the *mel* and the pitch of a tone, which is twice that of another tone, has twice the number of mels.

A tone having a frequency of 1000 cycles per second is arbitrarily defined as having a pitch of 1000 mels. Figure 10.1 shows the relation between the pitch scale in mels and the frequency scale in cycles per second.

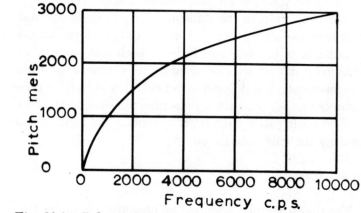

Fig. 10.1.   Relation between pitch in mels and frequency in cycles per second.

The pitch of a sound may depend on loudness. In the frequency range below 1000 cycles per second, the pitch of the sound decreases as the loudness increases. Between 1000 and 3000 cycles per second, the pitch is relatively independent of the loudness, while above 3000 cycles per second the pitch increases as the loudness is increased.

Pitch acuity is defined as the ability to detect differences in pitch (Δf). The minimum ratio of $\Delta f / f$, which can be detected in the intermediate range of frequency, is essentially constant; however, at low frequencies the value of Δf is essentially constant so that $\Delta f/f$ is larger. Good intonation is therefore more difficult for high-pitched than for low-pitched tones. Individuals differ in their pitch acuity, but it is possible to improve this with training.

Pitch recognition, which is the ability to recall or name a pitch, varies with the individual. Most individuals possess this ability to a greater extent when they are young because the ability is usually suppressed.

When a complex tone consisting of the frequencies 500, 600, 700 and 800 cycles per second is sounded, the ear hears it as the same pitch as a tone of 100 cycles per second, which is the fundamental of which the actual tones are harmonics. This is called a *subjective tone*. Subjective tones have an important application in small radio receivers. The speakers in such devices cannot effectively reproduce tones of frequencies below about 120 cycles per second but the musical tones of the lower frequencies have harmonics in the frequency range the speaker can reproduce and the presence of these harmonics conveys a pitch sensation to the listener of the fundamental of the tone.

The range of human voices extends from about 60 cycles per second for the lowest tone of a low bass to about 1300 cycles per second for the highest tone of a high soprano.

The maximum range of any voice, however, rarely exceeds about two octaves. Musical instruments have greater pitch ranges than the human voice. The range of the piano is from about 27 cycles per second to 4186 cycles per second, or slightly more than seven octaves. Some organs have an even greater range.

Tuning forks are often used as standard pitch sources

for tuning musical instruments. Pitch pipes are not so accurate as tuning forks but are often used because they are less expensive. Radio station WWV, operated by the U. S. Bureau of Standards, transmits two standard frequencies, 440 and 600 cycles per second. The international standard for *a* above middle *c* is 440 cycles per second.

10.3. *Quality*.

If the same note is sounded on a piano, violin, clarinet, flute or saxophone, the nature of the sounds produced are quite different, even though the fundamental frequency of the tone is the same in all instances. The distinguishing characteristic of these tones is called *quality* or *timbre*. This difference is due largely to the number and relative intensities of the harmonics or overtones, but is due partly to the rate of build up and decay of the tone.

The early mechanical phonographs did not reproduce frequencies above about 2000 cycles per second. Because of this, most of the harmonic content of tones from musical instruments was eliminated by the reproduction process, and as a result the tones lost their distinctive quality, causing one instrument to sound much like another. The human voice does not have much high harmonic content, which accounts for the fact that these early phonographs were quite successful for reproducing the human voice.

Using modern electronic amplifiers with filters, it is possible to make a recording of the sounds from a musical instrument, first covering only the frequency range up to 2000 cycles per second, then covering only the frequency range above 2000 cycles per second. If one listens to the playback of these recordings separately, it is surprising to note that, although by far the major portion of the sound energy is in the lower frequency band, the distinctive quality of the instrument is not apparent unless both bands of frequency are heard simultaneously.

In most instances all of the overtones of an instrument are harmonious—that is, they are exact multiples of the fundamental frequency. There are, however, some instances where some of the higher overtones are not harmonious.

In addition to the harmonic content of a sound the timbre or quality is partially determined by the rate of growth and decay of the sound. If the sounds from a violin and a flute are filtered so that most of the overtones are removed, a trained listener can still identify them because of this difference in growth and decay of the sound. A distinctive characteristic of a piano sound, for example, is a result of the fact that the string is struck with a hammer so the buildup of the sound is very rapid. The piano and harpsichord have similar string arrangments but the sound is quite different because the piano strings are struck and the harpsichord strings are plucked.

It is theoretically possible to duplicate the sounds of any instrument by combining the proper frequencies to produce the fundamental and the overtones with their proper relative intensities and controlling the build up and decay of the sound so that it follows the same rate of change as that of the instrument being simulated. This was first accomplished by Helmholz using tuning forks and resonators. The tuning forks were mounted on rubber pads so that their sound was suppressed until the resonators were opened. By controlling the sizes of the openings on the resonators, he could control the relative intensities of the overtones. Helmholz was able to demonstrate the principle by this method but, of course, he could sound only one note at a time.

With modern electronic equipment, it is possible to generate the desired tones and control the attack and decay of the tones to simulate any instrument. There are a number of electronic organs available now which quite effectively simulate the sounds of a pipe organ with all of the electronic circuits enclosed in a cabinet no larger than that of a small upright piano.

10.4. *Musical Intervals.*

Perhaps the earliest acoustic studies were carried out by Pythagoras, who lived from 572 to 497 B.C. Pythagoras studied the relations between tones which were pleasing to the ear when they were sounded together. He found that if a stretched string is divided by a movable bridge so that one segment is twice as long as the other, the combination of the tones emitted by the two segments of the string is most pleasing. By experimenting with different ratios of the lengths of the segments he showed that if the ratio could be expressed as the ratio of two small numbers the combination of the two tones was pleasing. Pythagoras had no means of measuring frequency but we now know that if the mass and tension of a string is held constant, the frequency of vibration varies inversely with the length of the string. Therefore, if the lengths of the segments of Pythagoras's string was 3/2, the ratio of the frequencies would be 2/3. If the tension of the string is increased, the frequencies emitted by the segments will increase but the ratios of the frequencies will remain the same.

We define the ratio of the frequencies of two tones as

their musical interval and the musical interval corresponding to a ratio of 1/2 is called an *octave*. The tones having
frequencies of 44 and 88 cycles per second have a musical
interval of one octave and likewise the tones having frequencies of 3000 and 6000 cycles per second have a musical interval of one octave.

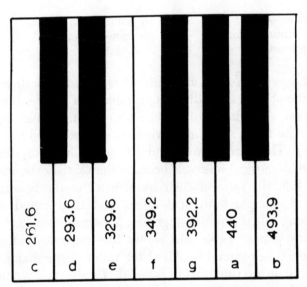

Fig. 10.2.   One octave of a piano keyboard starting at middle
c and including *a* at 440 cycles per second.

The musical interval is usually named from the position
of the two notes on a musical scale. Figure 10.2 shows
the keys for one octave on a piano. If we take the scale
of *c* major and label the notes from *c*, we have the white
keys of the piano. Table 10.1 shows the major and minor
scales with *c* as the keynote.

Table 10.1
The major and minor scales with *c* as the keynote.

Major scale

c		*g*	fifth
d	second	*a*	major sixth
e	major third	*b*	major seventh
f	fourth	*c′*	octave

Minor scale

c		*g*	fifth
d	second	*a♭*	minor sixth
e♭	minor third	*b♭*	minor seventh
f	fourth	*c′*	octave

The second, fourth and fifth are common to both scales but the second is hardly recognizable as a simple musical interval. We distinguish the scales as major and minor since the third, sixth and seventh are different on the two scales. The thirds, fourth and fifth are quite easily recognized as musical intervals because they satisfy the condition that the ratio of their frequencies to the frequency of *c* be simple numerical ratios, and they sound most pleasing when sounded together. You can easily test this on any piano. Table 10.2 shows the frequency ratio for these various intervals. All of the notes of the major diatonic scale are used except *b* and *d*. However, if we make *g* the keynote instead of *c*, *b* will be the third and *d* will be the fifth.

Musical intervals are measured in *cents* which are so defined that there are 1200 cents in an octave. A tone is 200 cents and a semitone is 100 cents in the tempered scale in use today. Referring to Fig. 10.2, the tones from any two adjacent keys on the piano have a musical interval of one semitone or 100 cents. Thus, the musical interval of c and $c\sharp$ is one semitone and likewise the musical interval of e and f is one semitone.

Table 10.2
Frequency ratios of the various musical intervals.

Notes.	Interval.	Frequency ratio.
$c - c'$	Octave	1:2
$c - g$	Fifth	2:3
$c - f$	Fourth	3:4
$c - e$	Major third	4:5
$c - e\flat$	Minor third	5:6
$c - a$	Major sixth	3:5
$c - a\flat$	Minor sixth	5:8

10.5. *Musical Scales.*

In all places where music has developed musical scales have evolved. Although there is an infinite variety of notes, our musical scale is divided into notes separated by a semitone and the frequencies between the notes are not used. It is apparently necessary to have the musical sounds separated by distinct steps rather than to glide continuously in frequency, in order to produce the desired emotional appeal to the listener. Musical scales evolved in the process of the development of music and the scales have been continuously revised as the music has developed. Since nature does not impose a standard form for musical scales, many different scales have been developed but they are similar in that the octave is taken as a unit and it is divided by the other notes.

Early Western music and the present-day music of the Chinese, Indians, Arabs and Turks is *homophonic*—that is, one part music. It is used for short pieces and as accompaniment for poetry. Only a single melody is involved in homophonic music. The great development in European music was due to the introduction of *polyphony* and *harmony*. Polyphony, or many part music, was apparently developed so that men and women or men and boys could sing together. Harmony is the integration of the parts to produce a pleasing effect. At first the melody was simply duplicated in the next higher or lower octave and later it was duplicated at an interval of a fourth or a fifth. Soon different melodies were woven together.

Most of the early musical developments were done in the monasteries, where life was rather dull, and experimenting with music provided an interesting pastime. Sometimes the melodies of popular songs were woven into the sacred melodies. In harmony, the primary concept is the single chord. As long as we deal with simple melodies, slight inaccuracies in the musical intervals are not noticeable because we hear only one note at a time. However, when we deal with harmonized chorales the intervals must be established with high precision or harsh discords will result. The Reformation stimulated the development of harmony because of the demand for singing harmonized chorales in the Protestant Churches.

The origin of our major and minor scales can be traced to the music of the Greeks. Music was important to the Greek people. Plato included it as an important part of education. Almost all of the Greek music has been lost but the contributions of the Greeks to the theory of music have been preserved in the writings of the followers of Pythagoras. There was, apparently, considerable development as early as 1200 B.C. The instrument concerned in this development was the tetrachord, an instrument with four strings. The tetrachord was tuned in two ways. The

tetrachord of Olympus may be considered in relation to
the white keys on the piano. Using the keys *a, f* and *e* or *e,
c* and *b* and adding the minor third below *e* we will have
e, c♯, c and *b*. This was the chromatic tetrachord. If we
take the minor third up from *b*, we can use the keys *e, d,
c* and *b*, which was the diatonic tetrachord.

If we let *T* be a musical interval approximately equal
to our present tone (200 cents) and *t* be an interval equal
to a semitone, three of the Greek scales or "modes" can be
shown in terms of these musical intervals in Table 10 3.

The scale was extended by adding a second tetrachord
with a tone interval between them. We then have the
three modes shown in Table 10.4. These three modes can
be approximately reproduced by using the white keys on
the piano as shown in Table 10.5. The Lydian system is
the same as our ordinary scale in *c* major. In the Greek
system a range of two octaves was eventually mapped into
seven modes which can be approximately represented
with the white keys on the piano by starting successively
with *b, c, d, e, f, g* and *a*, as shown in Table 10.6. If we
transpose these modes so that they all start with *c*, they
can be expressed as indicated in Table 10.7 where both
white and black keys must be used.

In each mode the scale consists of seven notes to the
octave and the seven notes divide the octave into seven
intervals of which five are larger and are called tones and
two are smaller and are called semitones. The composers
achieved variety by using different modes in which the
distribution of the tones and semitones are different.
These modes formed the basis of the homophonic music
of the early Christian Church.

With the introduction of harmony only two modes sur-
vived, the scale using the white keys of the piano begin-

ning with *c*, the old Greek Lydian, which is our modern major scale, and the scale using the white keys on the

Table 10.3
Three of the Greek modes expressed in terms of intervals of tones and semitones.

Species	Dorian			Phrygian			Lydian		
Diatonic	t	T	T	T	t	T	T	T	t
Chromatic	t	t	$T+t$	$T+t$	t	t	t	$T+t$	t
Enharmonic	$t/2$	$t/2$	$2T$	$2T$	$t/2$	$t/2$	$t/2$	$2T$	$t/2$

Table 10.4
The three modes with two tetrachords.

Dorian	t	T	T	T	t	T	T
Phrygian	T	t	T	T	T	t	T
Lydian	T	T	t	T	T	T	t

Table 10.5
The three modes expressed in terms of the white keys on the piano.

Dorian	*e*	*f*	*g*	*a*	*b*	*c*	*d*	*e'*
Phrygian	*d*	*e*	*f*	*g*	*a*	*b*	*c'*	*d'*
Lydian	*c*	*d*	*e*	*f*	*g*	*a*	*b*	*c'*

Table 10.6
The seven modes expressed in terms of the white keys on the piano.

Mixolydian	*b c d e f g a b*
Lydian	*c d e f g a b c'*
Phrygian	*d e f g a b c' d'*
Dorian	*e f g a b c' d' e'*
Hypolydian	*f g a b c' d' e' f'*
Hypophrygian	*g a b c' d' e' f' g'*
Aeolian	*a b c' d' e' f' g' a'*

Table 10.7

The seven modes from Table 10.6 tranpsosed so
that they all begin with c.

Mixolydian	c	d♭	e♭	f	g♭	a♭	b♭	c′
Lydian	c	d	e	f	g	a	b	c′
Phrygian	c	d	e♭	f	g	a	b♭	c′
Dorian	c	d♭	e♭	f	g	a♭	b♭	c′
Hypolydian	c	d	e	f♯	g	a	b	c′
Hypophrygian	c	d	e	f	g	a	b♭	c′
Aeolian	c	d	e♭	f	g	a♭	b♭	c′

piano starting with *a*, the old Greek Aeolian, which is our
modern minor scale.

Other scales have been developed, and as long as music
is homophonic there are many possibilities for scales, but
when harmony is used the number of scales that can be
used is greatly limited.

In Fig. 10.2 one octave of the piano keyboard is shown,
starting with middle *c*. The *a* in this octave has a frequency of 440 cycles per second and the frequencies of
the other notes in the octave are indicated in the figure.

11.

Musical Instruments

11.1. *Vibrating Bars.*

The simplest application of vibrating bars in musical instruments is the music box. The bars, usually made of brass or steel, are rigidly attached at one end and the bars are actuated by plucking the free ends with pins on a

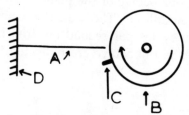

Fig. 11.1.   Mechanism of a simple music box.

rotating drum as indicated in Fig. 11.1, where A is the vibrating bar, B is the rotating drum, C is the pin that actuates the bar and D is the rigid frame to which the bars are attached. The frequency of vibration of the bar is determined by the elastic stiffness of the bar material, its mass per unit length and its length.

The bar vibrates in the bending modes and the fundamental mode is shown at A in Fig. 11.2. In this mode of

158

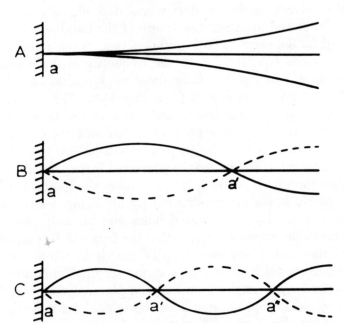

Fig. 11.2. The first three modes of vibration of a bar anchored at one end.

vibration there is a *node,* a point of no vibration, at the point of support and the amplitude of vibration increases continuously from this point out to the end of the bar. The standing wave is, therefore, ¼ wavelength for the length of the bar. The next higher mode of vibration of the bar is shown at *B* in Fig. 11.2. The free end will again be a point of maximum vibration and the point of support must be a node. In order to have a standing wave a second node must be established at *a'.* Since nodes must be separated by ½ wavelength, the length of the bar will be equal to ¾ wavelength so the wavelength of the standing wave will be ⅓ the wavelength of the fundamental. Since the velocity of the bending waves depends only on the stiffness and mass per unit length of the bar, the velocity will be the same in both instances, and because the frequency is equal to $f=v/\lambda$, the frequency of this second mode will be equal to three times the fundamental. The third mode is shown at C in Fig. 11.2. This mode is formed by the formation of nodes at a' and a'' in addition to the node at the support. The bar will then have a length of $\frac{5}{4}\ \lambda$ and the frequency will be five times the fundamental. The harmonics generated by such a bar will, therefore, have values 3, 5, 7, . . . times the fundamental.

Music boxes are constructed by arranging a series of bars to produce the desired notes and properly placing pins in the rotating drum so that the bars will be plucked in the proper sequence to play the desired tune. The drum is usually driven by a spring motor. Simple music boxes have a single drum and can play only one simple tune. More sophisticated music boxes can play more than one tune by using a different drum for each tune.

S_1 S_2

Fig. 11.3.   Vibrating bar of a xylophone indicating the fundamental mode of vibration.

The xylophone is another application of vibrating bars. In this application, the bars are usually made of wood and they are supported at two points as indicated in Fig. 11.3. The nodes for the fundamental will be at the points of support which are one-fourth the length of the bar from each end and the maximum amplitudes of vibration will be at the center and at the two ends of the bar. The Xylophone is played by striking the bars at the center with a small hammer.

The tuning fork is a variation on the kind of bar used in the xylophone except that the bar is bent to a U shape and a support is formed at the base of the U as indicated in Fig. 11.4. Because of the additional mass of the support,

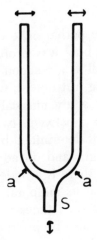

Fig. 11.4.   Tuning fork.

the nodes are near the bottom of the U at *a, a* and their position is determined largely by the mass of the portion between them. When the tuning fork vibrates, the stem S moves up and down and the prongs move as indicated in Fig. 11.4. The phase of vibration of the prongs is such that they alternately move towards and away from each other. The tuning fork is normally set in vibration by striking one of the prongs near the end with a small rubber hammer.

The tuning fork is not directly applied in musical instruments but it is a source of very constant frequency of practically pure tone and is, therefore, often used as a standard for tuning musical instruments. An interesting application of the tuning fork is incorporated in an electric watch in which a tiny tuning fork is used as a standard of time.

11.2. *Vibrating Strings*.

Vibrating strings are perhaps the earliest form of musical instrument. The bow was probably used as a musical instrument as well as a weapon by early man. The tetrachord, a four-string instrument, was used by Pythagoras (572-497 B.C.) to study musical sounds.

If a string is stretched between two fixed supports its frequency of vibration is determined by its tension, length and mass per unit length and, in addition to its fundamental mode in which there is a node at each end and a maximum of vibration at its midpoint, the harmonics corresponding to 2, 3, 4, 5 . . . times the frequency of the fundamental mode can also be excited. The intensity of sound emitted by a stretched string is very low unless the string is coupled to a sounding board or a sound box because of the small area of the string in contact with the air.

The quality of a note emitted by a stretched string depends on the number and relative intensities of the harmonics that are excited. There are three ways in which the quality of the note from a stretched string may be varied:

1. The point of attack may be varied. Strings on musical instruments are usually not excited at the midpoint. A point is selected that will cause the excitation of the

desired harmonics and also suppress harmonics that are
not desired.

2. The method of attack may be varied. Strings may
be bowed as in members of the violin family; they may be
plucked as in the harp, guitar, banjo and harpsichord;
or they may be struck as in the pianoforte.

3. The vibrating system to which the strings are
coupled may be varied to emphasize certain harmonics
and suppress others.

Instruments such as the harp, guitar, banjo and dulcimer may be considered to be descendents of the Greek
tetrachord. They are all plucked string instruments but
they differ in the kind of vibrating system to which the
strings are coupled and therefore they differ greatly in
the quality or timbre of the sound produced. Most of
these instruments have a small number of strings and
the required range of pitches is obtained by varying
the length of the vibrating portion of the strings by
stopping them at various positions with the fingers on
a fingerboard. The harp is an exception in that it uses
a large number of strings and the full lengths of the
strings are used at all times. All of these instruments are
played by plucking the strings with the fingers or with
a small pick held with the fingers.

The harpsichord is a plucked string instrument that
uses a mechanical system operated by keys to pluck the
strings. There is a separate string for each note and each
string is plucked by a mechanism operated by a key
similar to the keys on a piano. The keyboard of a harpsichord covers approximately five octaves. The simplest
harpsichords contain one set of strings which operate in
the same frequency range as that of the piano—i.e., the
a above middle *c* is 440 cycles per second. More elabo-

rate instruments may contain two or three additional sets of strings with one set tuned an octave higher than the standard set and one tuned an octave lower. Such instruments will have two keyboards with mechanical provision for a single key to operate either one or two plucking mechanisms. The harpsichord was an important keyboard instrument at the time that Bach was composing his music and much of his music was composed for the harpsichord.

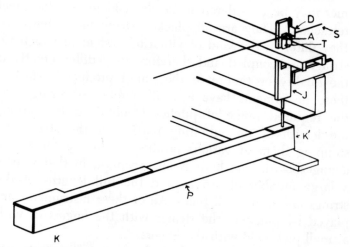

Fig. 11.5. Action mechanism of a harpsichord.

Figure 11.5 shows a diagram of an action mechanism of a harpsichord. The key is indicated at K with the pivot of the key at P. When the key is depressed at K, the end K' will be raised, which raises the jack J causing the leather plectrum A to pluck the string S. When the key is released, the jack drops back down and the plectrum slides back over the string because the tongue T

in which the plectrum is mounted deflects back against a weak spring to allow the plectrum to drop below the string again. When the jack drops back to its rest position a felt damper D rests on the string to stop its vibration. The loudness of the tone from the string is independent of the way in which the key is struck and is dependent only on the stiffness of the leather plectrum and the distance its edge projects beyond the string.

After the development of the pianoforte the harpsichord nearly became extinct, but it has recently become popular again, largely due to the efforts of Wanda Landowska.

If you are moderately skilled in wood craftsmanship you can build your own harpsichord by purchasing a kit from Zukermann Harpsichords, Inc., 115 Christopher St., New York, N. Y. 10014. The kit contains all of the special components necessary for construction of the instrument with drawings and a manual for construction and maintenance of the instrument. Figure 11.6 is a photograph of a Zukermann harpsichord constructed by the author.

The pianoforte was developed from the harpsichord. Its strings are not plucked, but are struck with a felt hammer using a mechanism designed so that when the key is struck the hammer swings against the string and then drops away from it so that the vibrations of the string will not be damped. When the key is released, a felt damper is allowed to contact the string to stop its vibration. The mechanism of the pianoforte permits the use of heavier strings at higher tension than can be used in the harpsichord. Multiple strings tuned in unison are used for some of the notes. The sounding board of the pianoforte is made of wood and the strings are coupled

Fig. 11.6. Photograph of a home-made harpsichord.

to the sounding board by means of a bridge, but the frame of the instrument is made of steel and the end supports of the strings are mounted in the steel frame. Because of this, the pianoforte will remain in tune for long periods. The harpsichord is made entirely of wood so that any change in the relative humidity will affect the tuning of the instrument.

The heavier strings of the pianoforte enable it to generate considerably higher sound levels than the harpsichord and the sound level of a note can be controlled by the force with which the key is struck. It is this feature which is responsible for the name of the instru-

ment that, in musical terminology means "soft loud."
The name, of course, is commonly shortened to "piano."

The strings of the pianoforte and the harpsichord are
wound on tuning pins at one end so that the instruments
can be tuned by turning the pins with a wrench to
tighten or loosen the strings. The sounding boards of
these instruments, which are coupled to the strings
through the bridges, radiate the sound to the surrounding air. The quality of the sound radiated is determined
to a large extent by the sounding boards.

In the violin family of instruments the strings are
coupled by means of a bridge to a hollow box in which
the vibration characteristics of the back and top as well
as that of the enclosed air volume determine the quality
or timbre of the sounds emitted. The strings are excited
by means of a bow which uses horse tail hair coated
with rosin.

The action of a bow on a string depends on the fact
that static friction is greater than moving or dynamic
friction. You can demonstrate this by attaching a rubber
band to a wood block and trying to slowly pull the block
along on a table top using the rubber band as a tow line.
The block will move in jerks because the block will remain stationary until the tension in the rubber band is
sufficient to overcome the static friction. When the block
starts to move, the dynamic friction will come into play
and, since it is less than the static friction, the block will
jump forward until the tension in the rubber band is
less than the dynamic friction force and the block will
stop until the tension in the rubber band has again
reached the value of the static friction force. The rosin
coated hair on the bow has an especially large difference

between the static and dynamic friction on the strings of the violin, so the bow is an effective device for exciting the strings to vibration.

The violin has four strings and the desired tones are obtained by stopping the strings with the fingers against a finger board to control the length of the vibrating segment of the string. The vibrations of the strings are coupled to the violin sound box by means of a bridge. The top of the box is made from spruce and the back is made from curly maple. Both the top and back are arched and scraped to the necessary thickness in order for them to have the required resonant frequencies.

We still do not understand all of the factors that determine the quality of sounds from a violin. The violins constructed by Guarnerius and Stradavarius are still superior to the best we are able to construct today.

Fig. 11.7.   Cross section of the sound box of a violin.

Figure 11.7 shows a cross section of a violin indicating the arching of the front and back, the bridge, the bass bar, the sound post, which couples the back to the front, and the purfling, which is a decorative inlay around the edges of the back and top and which may play a role in determining the tone quality of the instrument.

If you are a reasonably accomplished wood craftsman you can purchase the necessary materials and a manual for construction of a violin from the Ernst Heinrich Roth Co., 1729 Superior Avenue, Cleveland, Ohio 44114. They supply templates and instructions for constructing a copy of either the Stradavarius or the Guarnerius violins.

The viola and the cello are similar to the violin except that they are larger and are designed to cover lower pitch ranges.

The viol family preceded the violin. The viols are similar in appearance to the violins except that the back is not arched and they have six strings instead of four. The bass viol is still used as the lowest pitched, bowed string instrument in the orchestra. It is also used as a plucked string instrument, especially in dance orchestras.

11.3. *Vibrating Air Columns.*

The standing wave pattern of an air column was described briefly in Section 1.1 and illustrated in Fig. 1.8. We can consider the modes of vibration in the air column of a tube open at both ends and open at one end and closed at the other. At the open end of a tube the air is free to move so that it will always be a point of maximum vibration, or an *antinode*. The closed end of a tube is a point where the air is not free to move, so it will always be a node or a point of no motion of the air.

The fundamental mode of vibration of the air column in a tube open at both ends is shown in Fig. 11.8 a. Since the air column in a tube open at both ends will always vibrate with an antinode at both ends, the lowest mode of vibration will be that which has a node at the midpoint. The distance between an antinode and a node

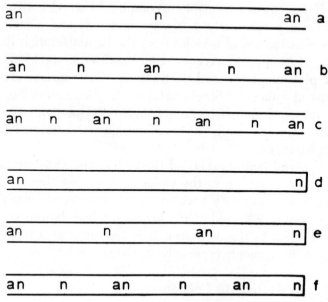

Fig. 11.8.   First three modes of vibration of an air column in an open tube and a closed tube.

is ¼ wavelength of the sound in air so the length of the tube will be ½ wavelength or $\lambda/2$. Since the velocity of sound in air is approximately 1000 feet per second and $\lambda = c/f$, the wavelength of sound in air for a frequency of 440 cycles per second is $1000/440 = 2.27$ feet. The length of the air column in the tube is $\lambda/2$ so the length of the tube for a fundamental of 440 cycles per second will be 1.13 feet.

The second mode of vibration of the open tube will require another node, and since there must still be an antinode at each end the positions of the nodes and anti-nodes will be as indicated in Fig. 11.8 b. The length of the tube for this condition will be 1λ, so the frequency will be the octave of the fundamental. It is obvious that the distribution of nodes and antinodes for the next higher mode of vibration will be as indicated in Fig. 11.8 c. The length of the air column will be $1\frac{1}{2}\lambda$ and the frequency will be three times the frequency of the fundamental mode. Such an air column will, therefore, vibrate with all harmonics of the fundamental.

When the tube is closed at one end and open at the other, the open end will be an antinode and the closed end will be a node. The lowest mode of vibration will therefore be that indicated in Fig. 11.8 d where the length of the air column is $\frac{1}{4}\lambda$. A closed tube will therefore have a fundamental frequency of vibration one-half that of an open tube of the same length. The next higher mode of vibration will require one more node, so the positions of nodes and antinodes will be that shown in Fig. 11.8 e. The air column will be $\frac{3}{4}\lambda$ so the frequency will be three times that of the fundamental mode. Figure 11.8 f shows the positions of the nodes and antinodes for the third mode of vibration of the air column in a closed tube. In this instance, the length of the tube is $5/4\lambda$, so the frequency of vibration will be five times that of the fundamental mode. Such an air column will therefore vibrate with the fundamental mode and the odd harmonics of the fundamental.

In order to utilize the vibrations of air columns in musical instruments it is necessary to have methods of excitation of the vibration. In the organ, this is usually done by means of a vibrating stream of air. The simplest means of exciting the vibrations of an air column is to blow across the neck of a bottle or jug. The neck is the open end of the air column, so when the column vibrates the air moves in and out of the opening and

causes the air stream to oscillate at the frequency of one
of the modes of the air column.

Fig. 11.9.   Actuating mechanism of a flue type organ pipe.

Figure 11.9 shows a flue type of organ pipe. The air
stream enters at A and passes across the edge at F. As
the air stream flows past the edge, it will oscillate at
frequencies determined by the natural modes of the
pipe.

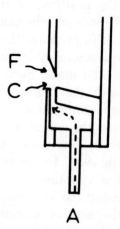

Fig. 11.10   Actuating mechanism of a stopped diapason type
organ pipe.

Figure 11.10 shows a diapason organ pipe. The air stream enters at *A* and passes through the slot *C* and impinges on the edge *F*. It will oscillate back and forth over *F* at frequencies corresponding to the natural modes of the air column. You can easily make a whistle which operates on the principle of the diapason pipe. Start with a piece of a willow branch about ½ inch in diameter and about 4 inches long with the ends cut as indicated in

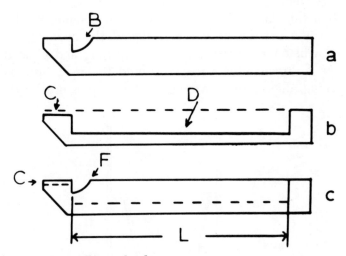

Fig. 11.11   Willow whistle.

Fig. 11.11 a. Cut a notch *B* with your knife and then tap the bark over the entire area with your knife handle until it loosens so that you can slip it off from the wood. With the bark removed, cut the wood as indicated in Fig. 11.11 b, removing a narrow strip at *C* and a wide strip at *D*. Now, if you slide the bark back on the wood you will have the configuration shown in Fig. 11.11 c where the channel *C* corresponds to *C* in Fig. 11.10 and

F corresponds to *F* in Fig. 11.10. Willow is the best wood for constructing such a whistle because of the ease with which the bark can be removed. Since this is a closed pipe, the wavelength of the fundamental will be four times the length *L* indicated in Fig. 11.11 c and the frequency $f = c/\lambda$ is approximately equal to 1000/4L cycles per second, where *L* is measured in feet, since the velocity of sound in air, *c*, is approximately 1000 feet per second.

If a thin flat reed is placed over an opening so that there is a gap between the reed and the opening, the reed will tend to flap against the opening and close it, when a stream of air passes behind the reed into the opening, due to the lowering of the air pressure by the

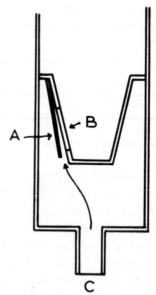

Fig. 11.12   Actuating mechanism of a reed pipe.

increase in velocity when the air flows between the reed and the flat area surrounding the opening. In Fig. 11.12, *A* is the reed, *B* is the position of the opening and *C* is the direction of the air stream. In use, the reed will vibrate at its natural frequency if there is no vibrating air column above it, but the presence of a resonant air column will tend to control the frequency at which the reed will vibrate.

11.4. *The Pipe Organ.*

The pipe organ is a keyboard instrument with an array of pipes that utilize the different types of vibrating air columns with the different methods of exciting their vibrations. The keys control valves, which in turn control the admission of air to the various pipes. The frequency of a given pipe is determined by its length and the timbre or quality is determined by the material of the pipe, its method of excitation and whether it is an open or a closed pipe. In addition, pipes are often arranged in sets with the fundamentals of some of the pipes in a set tuned to the harmonics of others, which gives an additional means of controlling the timbre.

The valves which control the air flow to the various pipes are operated electrically so that depressing a key simply closes an electric circuit. A large organ has a large number of *stops* available, which are simply switches to determine which pipes will be actuated by a given key. Such an organ will usually have two or three keyboards in addition to a foot pedal keyboard.

11.5. *Other Vibrating Air Column Musical Instruments.*

There are a number of musical instruments based on

vibrating air columns. These instruments have a single tube with means for changing the effective length of the vibrating air column. Instruments such as the flute, the recorder, the saxophone and the clarinet vary the effective length of the air column by use of holes in the side of the tube, which can be opened or closed by means of valves or the fingers. An open hole in the side of the tube allows free motion of the air at that point so the tube behaves as though the open end is at that position. The recorder is excited in a manner similar to that of the willow whistle shown in Fig. 11.11; the flute is excited by blowing across a blowing hole, called the *embouchure,* so the excitation is similar to that obtained when one blows across the neck of a bottle and the clarinet and saxophone are excited by means of a vibrating reed.

The brass instruments—such as the cornet, trombone, French horn and tuba—are metal horns. The effective length of the vibrating air columns is changed by inserting additional lengths of metal tube into the air column by means of valves or, in the case of the slide trombone, by changing the length of a telescoping tube. These instruments are excited by blowing between the compressed lips of the player while they are held against a mouthpiece at the end of the horn.

11.6. *Vibration of Stretched Membranes.*

The vibrations of stretched membranes are used in drums and the tambourine. The drum consists of a hollow cylinder of wood with a stretched membrane over both ends. The membrane over one end is struck and the vibrations are coupled from one membrane to the other by the enclosed air between them. The energy is passed back and forth between the membranes until it is dissipated by being radiated as sound. Some drums have a cord or wire, called a snare, stretched across the lower membrane. The membrane vibrates against the snare, giving it a special timbre. The drum is essentially a rhythm or beat instrument. The tambourine is a small drum having only one membrane. It also has metal discs or jingles at the side, which give a jingling sound in addition to the sound from the membrane when the membrane is struck.

The kettle drum has a single membrane stretched over the open end of a metal vessel shaped like a large kettle. The fundamental of the membrane has a node at the center and the frequency of the fundamental is inversely proportional to the diameter and inversely proportional to the square of its mass per unit area and directly proportional to the square root of its tension. The membrane is struck at a point about half way between the center and the circumference, which eliminates the fourth harmonic and weakens several others. Normally such a membrane vibrates with overtones that are not integral multiples of the fundamental and are therefore not harmonious. Provision is made in the kettle drum to vary the tension at different points around the circumference so the nonharmonious overtones can be tuned to harmony with the fundamental as the membrane is tuned. This permits use of the instrument in the orchestra to generate musical tones and the kettle drum player has a series of kettle drums of different size to provide the desired frequencies.

11.7. *Cymbals and Bells.*

Cymbals are circular plates depressed at the centers and provided with handles attached at the centers. They therefore vibrate with a node at the center. The operator uses two identical cymbals held in the two hands so that they can be struck together. Their modes of vibration are similar to membranes and the nodal lines are concentric circles due to the forced node at the center. A bell is a development of a plate cast in the familiar form. Bells are usually cast from an alloy consisting of 13 parts of copper to 4 parts of tin. The shape of bells was arrived at after centuries of experience and

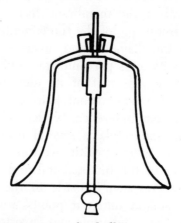

Fig. 11.13   Cross section of a bell.

is shown in Fig. 11.13. The largest bells are Great Paul at St. Paul's Cathedral in London, which weighs 16¾ tons, and Big Ben at Westminster in London, which weighs 13½ tons. The largest bell ever cast was the Great Bell of the Kremlin, which is estimated at 180 tons. Its height was over 19 feet. It was cast in 1734 and the clapper was pulled by 25 men on each side. This bell fell in 1737 and was partially buried in the fall and a portion was broken from it. It was raised in 1836 and placed on a pedestal.

11.8.   *Electric Instruments.*

Electronic organs have been developed, which utilize various electronic methods of developing frequencies with control of the generated harmonics of those frequencies. These electric signals are amplified and fed to loudspeakers. The quality of sound available from electric organs is limited only by the quality of reproduction of electronic systems. An important advantage of electronic organs is the fact that the entire system can be enclosed in a very small space compared to the space required by the array of pipes necessary to produce the corresponding sounds.

Many of the orchestra instruments can be equipped with electroacoustic transducers and the electronic system can be equipped with special filters to emphasize the desired harmonics and therefore serve as a means of controlling the timbre of the sound. Unfortunately most of the use of such devices is for the purpose of increasing the loudness of the sound from these instruments, and they are, therefore, contributing to the hearing hazard of people exposed to them.

12.

Room Acoustics

12.1 Introduction.

The early Greek theatres were not enclosed by walls. The seats were arranged in an approximate semicircle on a hillside with considerable rise from one row to the next row behind it. The audience in such a theatre received only the direct sound from the speaker or singer, so there were no acoustic problems if the sound level was sufficient to reach the entire audience.

When there was a performance at the theatre, practically all members of the community were in attendance, so there was no one to produce noise outside the theatre; and, of course, there were no aircraft, trains, motor vehicles or factories to generate noise to interfere with the performance.

The modern theatre, concert hall or auditorium is completely enclosed in a building. There are many sources of noise, such as jet aircraft and motor vehicles outside the building and ventilator fans and other activities in other rooms within the building, which must be shielded from the room if it is to be a desirable place to use as a theatre, concert hall or lecture auditorium.

Since such a room is completely enclosed by walls, ceiling and floor, any sound generated in the room will be reflected back and forth between these surfaces until all of the sound energy is absorbed and converted to heat. If a steady source of sound is started in a room, the level of sound will increase to a maximum steady state value; and if the source is stopped, the sound in the room will not cease abruptly but will gradually decrease to the point where it is inaudible.

When sound is reflected from any surface some fraction of the energy will be absorbed by the surface. The amount of the energy converted to heat at any reflection will be proportional to the intensity of the sound and a quantity called the *absorption coefficient, a.* The steady source of sound in a room will therefore cause the sound level in the room to increase until the rate of absorption of sound energy by the room surfaces is just equal to the rate of generation of sound energy by the source. If all surfaces in the room were perfect reflectors the sound level would continue to increase as long as the source continued to generate sound.

When the source of sound is stopped, the intensity gradually decreases because of the absorption of energy at the various surfaces. We define the sound which continues in the room after the source is stopped as *reverberation* and we define the time necessary for the sound intensity to decrease to one millionth (-60 dB) of its steady state value as the *reverberation time* of the room.

12.2. *Sound Insulation.*

In order to insulate a room from outside noises, the walls, ceiling and floor that surround the volume of the room must be of such construction that they will not effectively transmit the sound energy. If a homogeneous material is used in constructing a wall, the transmission loss through the wall will increase with the increase in the mass per unit area of the wall. Each time the mass per unit area is doubled the transmission loss will increase by about 6 dB, so the amount of sound insulation that can be achieved by simply increasing the mass of the walls is limited by practical considerations of weight of the structure.

In practice, the insulation is increased by utilizing the fact that reflection occurs when the sound passes from a medium having one value of acoustic impedance to a medium having a different acoustic impedance. If the boundaries of the room are made of multiple layers of material with the spaces between the layers filled with a material such as glass wool, multiple reflections will take place between the layers of structural material, and the padding between the layers will absorb the sound energy and convert it to heat.

12.3. *Reverberation.*

The acoustic characteristics of a concert hall have an important effect on the characteristics of the sounds that a listener in the audience hears. Some of the early European composers composed their music with a specific concert hall in mind so they took the characteristics of the concert hall as well as those of the instruments into account in developing their compositions. The early European concert halls were relatively small compared to present day concert halls and the architects learned by experience how to design them so that they would be acceptable for listening to musical performances.

When architects began to introduce innovations into the design of concert halls and auditoriums, the results were sometimes disastrous.

The pioneering work on the acoustics of auditoriums and concert halls was done by the late Wallace Sabine, Professor of Physics at Harvard University. He was called on to investigate the acoustics of an auditorium at Harvard University, and from this experience he developed an interest in architectural acoustics. He found that one of the important factors affecting the acoustic quality of a hall is its reverberation time. He also found that the reverberation time depends on the volume ot the hall and on the total absorption of sound by the surfaces of walls, ceiling and floor in the hall. Sabine defined a convenient unit for measuring absorption. Since an open window will allow all sound incident on the opening to pass out, he defined the unit of absorption as one square foot of open window. He could then define absorption coefficients of various building materials as the sound absorption of one square foot of the material compared to one open window unit (O.W.U.). Thus, a material that has an absorption coefficient, a, of 0.5 is one for which one square foot of the material will have an absorption of 0.5 O.W.U. If the total area of surfaces in a hall is known and the absorption coefficients of the materials are known, the total absorption of sound can be calculated since the total absorption A of a surface is equal to its area in square feet $\times$ a. Sabine derived a formula for calculating the reverberation time T of a hall as $T = 0.05\ V/A$, where V is the volume of the auditorium in cubic feet and A is the total absorption of all surfaces in the hall including that of the audience.

He found that the ideal reverberation time is a function of the use to which the hall is intended and to the volume of the hall.

In 1898 Sabine designed the acoustics of the Boston Concert Hall on the basis of studies of reverberation times of a number of halls recognized to have good acoustic characteristics, and this concert hall is recognized as having very desirable characteristics for music listening.

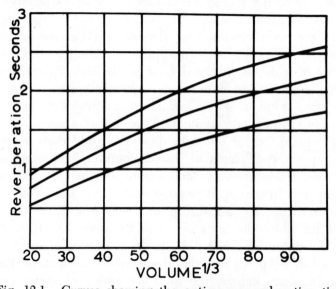

Fig. 12.1.   Curves showing the optimum reverberation time for halls as a function of the cube root of the volume for three kinds of program material.

If the reverberation time is too short, the hall sounds dead. If it is too long, speech syllables and rapidly played music become indistinct. Figure 12.1 shows reverberation times as a function of the cube root of the volume

$(V^{\frac{1}{3}})$ of the hall as determined from halls judged by experienced listeners to be good. These data show that the reverberation times should be longer for orchestral music than for speech, and they should be longer for choral and organ music than for orchestral music.

It is important that the performers as well as the listeners feel comfortable during the performance. If there is too little reverberation, the performers do not have a sense of ease and power in producing the tones, and it leads to frustration and fatigue. The performers apparently require higher levels of reverberation than the listening audience. This is achieved by surrounding the stage with hard reflecting surfaces. These surfaces also help to direct the sound out to the audience.

12.4. *Echoes.*

In an ideal auditorium, the reverberation builds up to fill the hall with a volume of sound. If a listener hears the first reflected sound at a time less than about 1/20 of a second after he hears the direct sound, the reflected sound will appear to reinforce the direct sound and the effect will be pleasing. If, however, he hears the first reflected sound more than 1/20 second after hearing the direct sound, he will hear it as an echo and it will interfere with his hearing of the direct sound. If the surfaces are broken up into small areas and the hall is not too large, the reverberation will build up smoothly and there will be no sensation of echoes. In Fig. 12.2, S is the position of a performer and X is the position of a listener. If D is the distance the direct sound travels to the listener and A + B is the distance the first reflected sound travels from a performer to the nearest reflecting surface

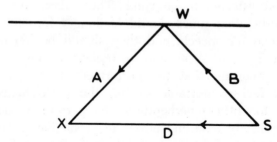

Fig. 12.2.   Paths of the direct sound and the first reflected
sound in a hall.

W and then to the listener, this reflected sound will reach
the listener more than 1/20 second later than the direct
sound if $A + B$ is more than about 50 feet greater than
D.

If the hall is small, there is little possibility of the first
reflected sound being received too late. The early European concert halls were small and the walls were irregu-
lar because of the decorations used in the architecture
of that period. Because of this, these halls were generally satisfactory if their reverberation times were correct.
Modern halls must be built considerably larger than
those early European halls because it is necessary to
accommodate larger audiences and the present standards
of comfort demand more space per person in the audience. Because of this, the walls and ceilings are so far
away from the performers that the time delay between
a listener hearing the direct sound and the first reflected
sound from the walls or ceiling will be considerably
greater than 1/20 second. In order to correct for this
deficiency, special reflectors, called clouds, are often
suspended below the ceiling and slightly forward of the

performers. These reflectors should have their spacing and height adjustable to produce the optimum relative level of the first reflection and have it reach the listeners at the optimum time.

There are several instances where halls have been constructed with a ceiling or walls having a concave shape. These surfaces will behave like the reflector shown in Fig. 4.4. If the radius of curvature of these surfaces is such that an image of a source of sound on the stage is formed at some point in the audience, a high intensity of sound will be focused at that point. If a symphony orchestra is playing on the stage, a listener may hear some instrument in the orchestra at a very high level compared with the rest of the orchestra because he is at the focal point of the mirror for that instrument. Examples of such halls are Albert Hall in London and a concert hall on the campus of the University of Illinois. These halls had to be modified considerably before they were satisfactory for use as concert halls.

In order to construct a good concert hall or auditorium, it is necessary to have close cooperation between the architect, the acoustician and the musicians. It is possible for an experienced acoustician to anticipate most of the problems in the design of a hall but, with our present state of knowledge, it is not possible to design the perfect hall. The design should be such that there is a chance to vary some of the parameters after the hall is completed. This adjustment or tuning of the hall must be done with the cooperation and help of musicians or other skilled listeners. If the hall is to be used for different kinds of programs, the architect should provide means for obtaining different reverberation times so that the optimum reverberation time for each kind of program can be obtained. An architect should employ the services of a competent acoustician from the beginning of the design of the hall. If he will heed the advice of the acoustician he can avoid the design features that are apt to lead to acoustic problems. A large concert hall represents a difficult problem under the most favorable circumstances, and the sponsors, the builder and the architect should reconcile themselves to the fact that listening perfection will not be achieved on the first try and they must be prepared to make changes after the building is completed in order to correct deficiencies.

Many studies are being carried out, some with the use of scale models with frequencies in a range such that the ratio of the sound wavelengths to the dimensions of the model are the same as they would be in the finished auditorium. It is hoped that the architectural acousticians will eventually learn enough about the problem in order for a hall to be designed in such a way that when it is completed it will have the optimum acoustic characteristics for its intended use.

12.5. *Sound Reinforcement.*

There are many instances when the sound level produced by a speaker or singer is not adequate to supply the desired level of sound to the listener. The logical solution of this problem is to utilize electronic amplification of the sound. At first sight the problem appears simple. All that is needed is to have a good microphone to pick up the sound from the speaker or artist, amplify the signal from the microphone with a good amplifier and supply the amplified signal to a good loudspeaker.

The usual results obtained by this process are disappointing. There is a tendency for the system to howl if the gain of the amplifier is turned up sufficiently to attain an adequate sound level and even if it is not turned up to the level where it will howl there is usually more distortion in the signal than would be expected from the characteristics of the components of the system.

In order to understand why this problem is more complex than it first appears, let us consider the system that includes a microphone, an amplifier, a loudspeaker and a room. If a sound is generated at the face of the microphone, the signal will be amplified and a higher level signal will be emitted by the speaker. This sound will be reflected around the surfaces of the room and some part of it will return to the face of the microphone. If the sound that returns to the microphone is out of phase with the exciting sound, the two sounds will tend to cancel. If the reverberent sound is returned in phase with the exciting sound we say that there is *positive feedback* in the system. When positive feedback exists in a system involving an amplifier the system will oscillate if the gain of the amplifier is sufficient. When the reverberant sound is returned out of phase with the exciting sound we say that there is *negative feedback*.

When such a sound reinforcing system is used in a room there will be a certain limited number of frequencies for which the feedback in the system will be positive. Fortunately, we can introduce negative feedback electronically in an amplifier and control the frequencies at which the positive feedback from the room takes place by a scheme developed by Dr. C. P. Boner.

Using this method, the sound reinforcement problem is approached by turning up the gain of the amplifier until the first resonance takes place, as is evidenced by the system starting to howl. The frequency of the howl is determined and a tuned negative feedback loop is inserted in the amplifier circuit, giving just enough feedback to stop the oscillation. The amplifier gain is then further turned up until the next frequency of oscillation is established and another negative feedback loop is inserted to prevent that oscillation. This process is continued until all of the frequencies at which the room and reinforcement system will oscillate have been neutralized by suitable negative feedback loops in the amplifier. The acoustic reinforcement system in the Astrodome is an example of what can be achieved when these principles are applied.

A method of sound reinforcement that has had considerable application in large rooms with low ceilings uses several small speakers in the ceiling distributed fairly uniformly around the room. These speakers are all driven from a single amplifier. Such a system avoids the use of a single high-intensity source, so that the normal reverberation pattern of the room is not excited and there is, therefore, little tendency for the system to howl. This system is commonly used at meetings and conventions where the listeners are primarily interested in being able to hear and understand the speakers and they are not particularly disturbed by the fact that the speaker is in front of them and they may hear the sound of his voice from the ceiling above and behind.

With improved technology in sound reinforcing systems we may some day see the use of such systems to help to solve the problem of the large concert hall. To achieve this it will be necessary to solve some psychological problems as well as technical ones. There is a

tendency for many people to feel that use of sound reinforcing is somehow cheating. The use of sound reinforcing systems in places such as nightclubs, where the
sounds are often amplified to the point where pain is
induced, contributes to this attitude.

13.

Underwater Sound

13.1. *Introduction.*

The field of underwater acoustics is one that developed originally because of its importance in military application. During World War I, in the search for methods of detecting submarines, it was soon learned that sound is the only form of energy propagated over appreciable distances in sea water and the propellers on a vehicle such as a submarine generate noise, which can be used as a means of detection.

Research on underwater sound was continued after World War I and by the beginning of World War II the state of underwater sound technology was considerably advanced. Because of the submarine menace during World War II, great advances were made in underwater sound transducer development and in our knowledge of the factors that influence the propagation of sound in the ocean. The use of underwater sound in navigation and ranging was given the acronym Sonar, which is derived from "sound navigation and ranging."

Since World War II the nonmilitary applications of underwater sound have received increased attention. It has been possible to draw on the large fund of information developed in antisubmarine warfare research for these applications. Submarines are detected by listening for the noise generated by their propellers or by transmitting a pulse of sound and listening for the echo. This latter method is called *echo ranging*, because the time between the transmission of the sound pulse and the receipt of the echo measures the range to the submarine. This same technique is used in the *fathometer* by projecting the sound pulse vertically downward to determine the distance from the keel of the ship to the ocean bottom. Fishermen now apply echo ranging for detecting schools of fish, oceanographers use it to study marine life and to determine bottom contours in the ocean. Archeologists use underwater sound to locate old sunken ships and even old centers of civilization which have sunk into the sea. Marine geologists use underwater sound to study the layers below the bottom of the oceans and the marine biologists use it to study the habits of marine animals.

13.2. *Generation and Detection of Underwater Sound.*

In Section 1.4 it was stated that the acoustic impedance of air is 42 grams per square centimeter per second and that of water is 150,000 grams per square centimeter per second. Therefore, for a given intensity of sound, the particle velocity in air will be 3570 times as great as the particle velocity in water and the sound pressure in water will be 60 times as great as the sound pressure in air.

Because of the great difference between air and water as media for sound propagation, transducers for underwater sound generation and detection must be quite different from those used in air. The active faces of the transducers must exert larger forces on the medium and the amplitudes of motion will be much less. In addition, the underwater sound transducer must be sealed against the sea water and it must have the necessary strength to withstand the pressure of the water at great depths. If a transducer has a face area of ten square inches and it is lowered to a depth of 2000 feet, it will have to support a load of 10,000 pounds on the face.

An underwater sound signal, which is not intended to be directional, is usually generated in a ring-type transducer, which may be either magnetostrictive or ceramic. Figure 6.10 shows the method of winding a magnetostrictive ring stack. If the transducer is intended to be directional in one plane and nondirectional in the other, a line of rings may be used. If the transducer is to be directive in both the horizontal and vertical planes, the elements are arranged in a plane array with the dimensions of the array relative to the wavelength determined by the desired directivity. If the same directivity is desired in both planes, the array will be symmetrical but if different directivities are desired in the two planes, the length of the array will be greatest in the plane of greatest directivity. In plane arrays, the elements are usually ceramic discs or magnetostrictive hairpin stacks similar to the one shown in Fig. 6.9. These elements are coupled to the water through a face of rubber, which has the same acoustic impedance as water. The elements must be strong enough to support the water pressure to which the transducer is subjected.

When it is desired to generate a given sound level, I_1, at some range, r_1, Eq. (3) from Section 4.1, namely:

$$I_1/I_0 = 1/r_1^2,$$

holds also for underwater sound.

The intensity I_0 is determined by the total electrical power input to the transducer, the efficiency of the transducer and its directivity factor. Most underwater sound applications use pulsed signals so that the electrical power, which can be supplied to the transducer during the pulses, is much greater than could be supplied to it on a continuous basis.

The total acoustic power emitted by the transducer must pass through its face into the water. The intensity of sound is equal to the power per unit area and the sound pressure is equal to the square root of the product of the intensity and the acoustic impedance of the water. In air acoustics, the only limitation on the sound intensity that can be generated by a source is the power-handling capability of the source. In water, however, there is another limitation due to the fact that during the negative half cycles of the sound pressure signal, the sound pressure is subtracted from the ambient water pressure, and when the sound pressure becomes approximately equal to the ambient pressure, water vapor will be formed during the negative half cycles. This effect, called *acoustic cavitation,* manifests itself as a cloud of tiny bubbles in front of the transducer. These bubbles scatter and absorb the sound, so the cavitation level is the practical limit of power level for the sound-generating transducer. As the depth of operation of the transducer is increased, the ambient pressure increases so the power level at which cavitation appears will increase. Near the surface of the ocean, cavitation will appear at about two watts acoustic power per square inch of transducer face.

The electroacoustic transducers are reciprocal devices. In underwater acoustics, sound is often generated by means of non-reciprocal types such as the underwater spark and explosive sources.

The same types of electroacoustic transducers as are used for sound generation can be used for sound detection and they will have the same directivity when used as receivers as they have when used as projectors. Transducers used as receivers do not need to meet the large power handling requirements of sound projectors but in all other respects the requirements are the same.

There are instances when a sound receiver is used to determine the direction to a sound source or its *bearing* in addition to detection of the sound. If the transducer has a high directivity, it can be used in the *searchlight* mode; that is, the transducer is oriented to the direction in which the signal level is a maximum. The source of sound will be in the direction in which the acoustic axis of the receiving transducer points. This is a rather crude method of determining bearing and can be used only when the source of sound does not fluctuate in level.

The method of bearing measurement most commonly used is based on the use of sum and difference patterns of the transducer. If we connect the elements in the two halves of the transducer in phase opposition, as in Fig. 13.1 A, the signal level as a function of bearing is indicated in Fig. 13.1 B. For signals received on the acoustic axis of the transducer, the response of the two halves exactly cancel each other and the response is zero. However, there are two equal lobes of sensitivity, one to the left and the other to the right of the axis.

The normal sum pattern produced when the entire transducer is connected in phase, as in Fig. 13.2 A, is indicated in Fig. 13.2 B. If the ratio of the signal on the difference pattern to that on the sum pattern at each

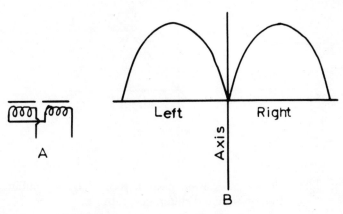

Fig. 13.1.   Difference pattern of a hydrophone.

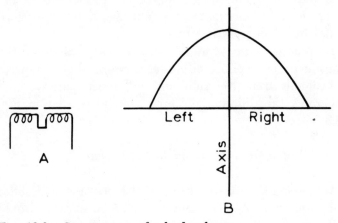

Fig. 13.2.   Sum pattern of a hydrophone.

bearing is plotted as a function of the bearing angle, the curve in Fig. 13.3 is obtained. The signals arriving from the left of the acoustic axis on the difference pattern can be distinguished from those arriving from the right, because in one case the difference pattern signal leads the

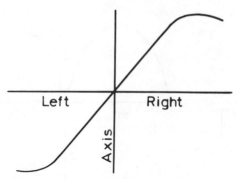

Fig. 13.3.   Bearing deviation pattern formed by a combination of sum and difference patterns.

sum pattern signal by 90 degrees in phase and in the other the difference pattern signal lags the sum pattern signal by 90 degrees in phase.

In practice, the terminals of the two halves of the transducer are connected to a network that compares the signals from the sum and difference patterns and generates the differential signal voltage plotted in Fig. 13.3.*

13.3.   *Propagation of Sound in the Ocean.*

When sound is transmitted in the ocean it will suffer the same loss of level due to divergence as it would suffer in any other medium. That is, it will suffer a loss of 6 dB in level each time the distance is doubled, as indicated in Section 4.1. The sound is also attenuated in the

---

* For a more complete explanation of the formation of sum and difference patterns of transducers and their use in bearing determination, the reader is referred to *Underwater Acoustics Handbook II* by Vernon M. Albers, Pennsylvania State University Press, 1965, Chapters 13 and 19.

water and the attenuation is a function of frequency. Table 13.1 shows how the attenuation of sound varies with frequency in sea water. The attenuation in fresh water is somewhat less than that in sea water because the salts in the sea water contribute to its attenuation of sound.

The velocity of sound in the ocean is approximately 5000 feet per second—that is, about five times that in air. If the velocity were constant throughout the ocean, it would be relatively easy to compute the intensity of sound at some distant point due to a certain sound level generated by a sound projector. The velocity of sound varies with the temperature, the pressure and the salinity of the water. The normal salinity or salt content is 35 parts per thousand, the pressure increases by approxi-

Table 13.1.
Sound attenuation in sea water as a function of frequency.

Frequency c.p.s	Attenuation dB per 1000 yds.
100	0.0001
1000	0.0045
5000	0.1
10,000	0.4
50,000	9.0
100,000	25.0
500,000	90.0
1,000,000	250.0

mately 0.5 pounds per square inch for each foot increase in depth, but the way in which the temperature changes with depth varies from point to point in the ocean. The sound velocity increases with temperature, pressure and salinity according to the empirical formula

$$c = 4625 + 7.68(t-32) - 0.0376(t-32)^2 + 3.35S + 0.018d, \qquad (13.1)$$

where c is the velocity of sound in feet per second, t is the temperature in degrees F., S is the salinity in parts per thousand and d is the depth in feet.

In the air, there is continuous mixing due to the wind, so the layering of air by temperature gradients is minimized. In the ocean, the wind and waves at the surface produce mixing in the surface layer but this layer extends in depth only to about 25 to 100 feet, depending on the speed and duration of the wind. As a result, layers with relatively sharp boundaries and considerable temperature difference from adjacent layers usually exist in the ocean. The boundaries of these layers are commonly called *thermoclines*. The depths of the boundaries of these layers are fairly uniform over quite large areas but there are times when waves in the boundaries of the layers exist. These waves are called *internal waves*.

The temperature of the water at the surface is determined by the heat exchange between the water and the atmosphere. When there is mixing at the surface, the mixed surface layer will have little or no temperature variation. Such a layer is called an *isothermal* layer. Throughout this layer there will be a slight increase of sound velocity with depth due to the increase in pressure. Below this depth, the temperature will usually decrease, and because the velocity is affected to a greater extent by temperature variations than by pressure variations, the velocity will decrease with depth. Below a depth of about 2000 feet the temperature becomes essentially constant so the velocity will then be controlled by the pressure and will increase with depth.

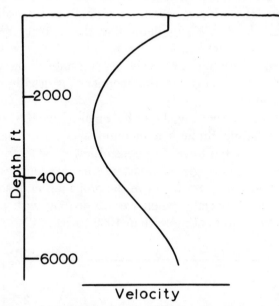

Fig. 13.4.   Idealized velocity profile in the ocean.

Figure 13.4 shows an idealized plot of sound velocity as a function of depth. Such a plot can be obtained by lowering an instrument called a *sound velocimeter* in the ocean. The velocimeter produces two signals, one a measure of the sound velocity and the other a measure of depth. These signals are fed to an automatic plotter, which plots the velocity versus depth curve. In most instances the sound velocity varies with depth in a more complex manner than that indicated in Fig. 13.4.

We can derive the sound velocity versus depth curve if we know how the temperature and salinity vary with depth. An instrument called a *bathythermograph* plots the temperature versus depth and a *salinometer* plots the salinity as a function of depth when they are lowered in the ocean. If we know how the temperature and salinity vary with depth we can determine the velocity versus depth curve by application of Eq. (13.1). Since a large number of calculations are necessary to establish the velocity versus depth curve in this manner it is usually done with a computer.

In Section 4.3 we saw how, by Snell's law, the sound rays propagating through a medium where the velocity varies are refracted toward regions of lower velocity. This effect does not have great importance in air acoustics but it is the most important factor in the propagation of sound in the ocean, especially when we are dealing with propagation over ranges of more than 1000 yards.

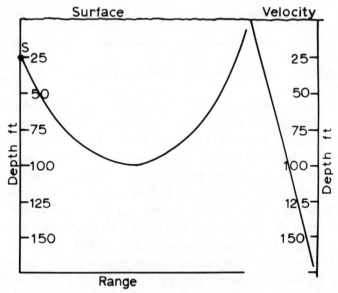

Fig. 13.5.   Velocity profile and sound refraction in a positive temperature gradient.

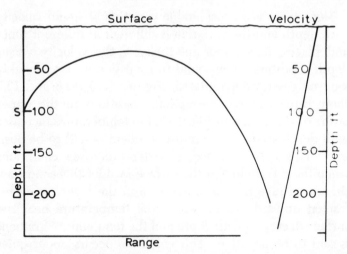

Fig. 13.6.   Velocity profile and sound refraction in a negative
              temperature gradient.

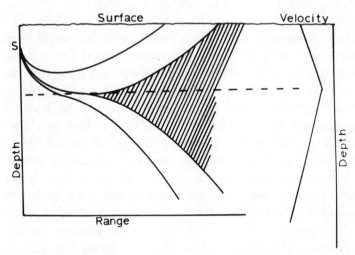

Fig. 13.7.   Formation of a shadow zone when a positive tem-
              perature gradient overlies a negative temperature
              gradient.

Since there is a considerable variation of sound velocity with depth and the variation is different at different times and places, it is important to know the velocity versus depth condition when sound transmission over reasonably long ranges is contemplated. Figures 13.5, 13.6 and 13.7 show three common refraction conditions in the upper layers of the ocean. In Fig. 13.5, the temperature increases with depth so the temperature gradient is said to be positive. This condition occurs under quiet sea conditions when there is cooling at the surface. Under this condition, the sound rays from a source at S in the layer are all refracted upward. In Fig. 13.6, the temperature near the surface decreases with depth and the temperature gradient is said to be negative. This condition occurs under quiet sea conditions when there is heating at the surface. The sound rays from a source at S in the layer are all refracted downward. In Fig. 13.7, we have a relatively common condition where a positive temperature gradient lies above a negative one. All rays above a certain angle from the source will be refracted upward toward the surface, but rays below this angle will suffer upward refraction before reaching the depth of maximum velocity and below that depth they will be refracted downward. This leaves a region indicated by the shaded area into which no direct rays from the source can penetrate. We call such a region a *shadow zone*. Although no direct rays from the source enter the shadow zone, the intensity of sound there is not zero because some surface reflected and some scattered sound does enter, but there is a large reduction in intensity in passing across the boundary of the shadow zone.

Figure 13.8 shows a less common thermal condition. Here there is a velocity minimum in the upper layer, and since all rays are refracted toward the depth of minimum

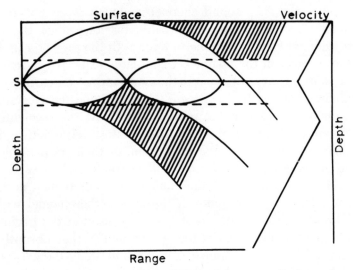

Fig. 13.8. Formation of a shallow water sound channel.

velocity a sound channel will be formed at this depth. Because there is no divergence in the vertical plane in the sound channel, the transmission loss due to divergence will be three dB each time the range is doubled instead of six dB. A sound channel such as this is very desirable for long-range transmission at shallow depth.

In Fig. 13.4 there is a velocity minimum when we reach the depth where temperature ceases to be the controlling factor in the velocity gradient and pressure becomes the controlling factor. This velocity minimum usually occurs at depths of 2000 to 4000 feet. This depth region is a sound channel because it is a depth of minimum velocity. It is called the deep sound channel and it is always present in the deep ocean. Like the shallow-water sound channel, the transmission loss due to divergence is three dB per double distance.

Since the deep sound channel is always present in the deep ocean it is always available for long-range transmission when it is possible to have both the projector and the receiver in the channel. Using large explosive sources of sound, ranges of several thousand miles can easily be obtained at low frequencies if the source and receiver are both on the axis of the sound channel. At the receiving station, the rays that suffer the greatest refraction and therefore spend the greatest portion of their transit time at the edges of the channel will arrive first because the velocity on the axis of the channel is lower than it is at the edges. An explosion on the axis of the channel will be received as a signal that has a low level at the beginning, gradually increases to a maximum and then abruptly stops. The time of transit of the sound along the axis of the channel is equal to the interval from the time of the explosion until the time when the signal stops at the receiver.

13.4. *Studies of Marine Animals.*

In the early use of the fathometer, weak echoes were obtained in large areas of the ocean from depths considerably less than the depth to the bottom. The levels from which these echoes were returned have a diurnal variation, being shallower at night than in the daytime. These layers are called the deep scattering layers and it has been demonstrated that they consist of living marine organisms.

Attempts to capture these organisms in nets have been ineffective since their concentration is very low. Studies of sound reflected from individual members of the deep scattering layers indicate that at least some of them have swim bladders and there is approximately one scatterer in about 1000 cubic yards of ocean. Deep scattering layers are present in most areas of the ocean and acoustic methods are about the only means we have to study them.

There has been considerable study of fish by echo ranging methods. Many of these studies have been for the purpose of obtaining information for the design of Sonars for fishermen. Some of these studies, however, have been for the purpose of learning the habits of fish, especially their schooling habits. Dr. Weston, of the Admiralty Research Laboratory in England, has used an advanced form of Sonar to study their schooling habits.

There is evidence that the fish use low-frequency sound to orient themselves in schools. All fish have ears which are similar to human ears although they are more primitive. Studies of the hearing sensitivity indicate that most fish have their maximum of hearing sensitivity at about 500 cycles per second.

Many fish have a second sound sensitive system in addition to the ears. This system is called a lateral line. It extends along each side of the body and portions of it are around the head of the fish. The lateral line consists of a line of hair-like cells that project into a channel on the surface of the body of the fish. Little is known about the sensitivity and frequency range of the lateral line but it is believed that it is a system which is sensitive to particle velocity in the medium and functions at very low frequencies. It is believed by some that fish use the lateral line to detect the turbulence produced by the neighboring fish, and that it is, therefore, used by the fish to position themselves in schools.

There is an important difference between a pressure sensitive and a particle velocity sensitive device as was indicated in Section 6.3. The sound pressure is a *scalar* quantity, which means that there is no direction indicated by the pressure. The particle velocity, on the other hand, is a *vector* quantity, which means that it has direction as well as magnitude and the particle velocity direction is always parallel to the direction of propagation of the sound.

Since the lateral line is a particle velocity detector, it is believed by some observers that the fish are able to detect predators by the particle velocity at very low frequency produced by the swimming motion of the predator. Since such a system is directional, the fish can sense the direction from which the predator is approaching.

There has been considerable study of the sounds produced by porpoises and whales. There is evidence that these animals communicate by means of sound. The sounds emitted cover a wide range of frequency, extending above 100 kc. In addition to possible communication with sound, these animals use sound in a manner similar to SoNAR for navigation and for finding food. Instead of an array of elements for their transducer, they have a fatty element called the *melon* on the front of the head, which apparently acts as an acoustic lens. Tests conducted with porpoises in tanks indicate that they are very adept at steering around obstacles, which they detect and locate by continuously transmitting pulses of sound. As they approach the obstacle the rate of transmission of the sound pulses increases.

13.5. *Bottom Topography.*

The fathometer, which was developed by Dr. Harvey Hayes in 1921, was one of the earliest practical applica-

tions of underwater sound. The obvious application of
the fathometer is as a means of being certain that there
is sufficient depth of water for the ship to traverse. The
fathometer is also useful for mapping the contours of the
ocean bottom. On land, we do much of our practical
navigation by making use of landmarks. The landmarks
in the ocean are the hills and valleys on the ocean bottom.
When accurate maps are available, which show the bottom contours in all parts of the ocean, it should be possible to navigate by means of a fathometer record. Since
about 70 percent of the earth's surface is covered by the
oceans, it will be a long time before we have complete
detailed maps of the bottoms of all of the oceans.

The ordinary fathometer is not capable of making precise determinations of depth because the time of return
of the bottom echo relative to the time of transmission of
the sound pulse must be determined to within one millisecond if the depth is to be determined to within five
feet, since the velocity of sound in the ocean is about five
feet per millisecond. The precision depth sounder is a
refinement of the fathometer for determining depth with
high precision. In order to achieve the necessary precision, a recorder is used. It has a single-turn helical blade
that turns against a knife blade with a continuous chart
of chemically impregnated paper moving between them.
The point of contact between the helix and the blade
starts at one edge of the paper when the first pulse is
transmitted and moves across the paper as the helix rotates. When the helix completes one rotation, it immediately starts the next and a second pulse is transmitted.
Each time the helix enters the edge of the chart another
pulse is transmitted and one period of rotation of the
helix corresponds to a depth of 400 fathoms. In deep

water, the helix may make several revolutions between the transmission of a pulse and the return of its echo, so a switching system is provided to determine within which 400-fathom interval the bottom is located. When an echo is received an electric current flows from the helix to the knife blade through the chemically treated paper causing the paper to darken. The precision depth sounder can determine depths of 3000 fathoms to an accuracy of about one fathom.

The precision depth sounder is· capable of measuring depth to high precision by transmitting a signal vertically downward and recording the returned echo from the bottom. It will not indicate the details of rock structures or the presence of sunken ships on the bottom. Another special kind of SoNAR, called the side scan SoNAR, is used for this kind of study.

If an acoustic signal is transmitted parallel to the bottom from a source near the bottom, a return signal is scattered from the bottom much as light from a flashlight is scattered back to an observer if he points the flashlight along a rough surface. The side scan SoNAR is mounted in a body, which is towed so that it travels near the bottom. Pulsed signals are transmitted in vertical fan-shaped patterns from each side of the towed body and the scattered return is recorded in a manner similar to the recording of the echoes with the precision depth sounder. Any rock or other structure that projects up from the bottom will cast a shadow so the record will be dark due to the sound scattered back from the bottom except in the shadow areas. The records are made by periodically transmitting acoustic signals and recording the received scattered signal on a chemically treated paper, which moves continuously so that a picture is developed from a series of lines much like the picture on a television tube is developed. From such a record, it is possible to estimate the shapes of any structures on the bottom.

13.6. *Marine Geology.*

Geology is a science that deals with the history of the earth as recorded in the rocks and sediments. On land, geology is studied by drilling cores and by examining the strata at the edges of mountains. It is also studied acoustically by measuring the propagation of seismic waves transmitted through the various rock layers and reflected from them when charges are exploded at fixed locations and detected by seismic detectors.

The oceans cover a large part of the surface of the earth and the structure of the areas under the ocean is as important to an understanding of the geology of the earth as that under the land areas. By use of coring devices, the oceanographers have been able to study the sediments in the upper few feet of the ocean bottom in some limited areas.

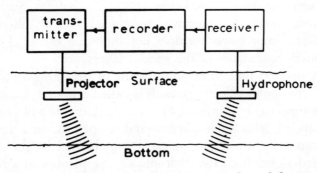

Fig. 13.9. System for indicating layers in the sub-bottom.

Underwater acoustics has provided a new tool, which makes possible studies to much greater depths in the sub-bottom than is possible by means of core samples. One system, shown in Fig. 13.9, consists of a projector such as an underwater spark, which projects a high-level, lowfrequency signal that penetrates the sea floor to great depths. When the sound waves cross a boundary of different acoustic impedance, part of the sound energy is reflected and part is refracted to greater depths. The reflected sound is received on the hydrophone; the signal is processed in the receiver and sent to the recorder. The recorder has a continuously moving paper chart similar to that in the precision echo sounder and the recording stylus moves from left to right across the paper during the time interval between transmitted pulses. The stylus darkens the trace by an amount that depends on the strength of the signal received by the hydrophone. The system shown in Fig. 13.9 is towed by a ship and as the ship travels along, the projector transmits signals periodically and a picture of the sediment and rock layers below the bottom is developed on the chart by the recorder. It is possible to record the profiles of rock layers as much as two miles below the ocean bottom. Such a system shows the interfaces between layers of different acoustic impedance but does not show the values of the acoustic impedance of the various layers.

A method that shows the depths of the various layers and the sound velocity in each is called seismic refraction shooting and is shown in Fig. 13.10. A number of buoys equipped with hydrophones and amplifiers, and radio transmitters whose output can be modulated by the amplified signal from the hydrophone, are planted in a row

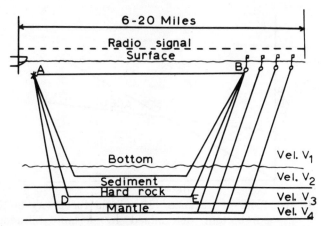

Fig. 13.10. Seismic refraction shooting to measure sound velocity in layers in the sub-bottom.

by a ship, which travels in a straight line and drops explosive charges overboard at intervals.

At short ranges, the signal due to the sound traveling the direct path AB will be received first by the radio receiver on the ship. As the range increases, sound traveling through the more dense layers where the velocity is greater will arrive first. If the ship travels at a uniform speed, and the times of arrival are plotted as a function of range, as in Fig. 13.11, a series of straight lines will result, each of which represents the propagation of sound in a stratum in the sub-bottom. The slopes of the lines give the velocity of sound in the strata (velocity = distance/time for any point on the line) and the depths of the strata can be determined by the intercepts on the time axis where the range is equal to zero because the intercept gives the time it would take for the sound to

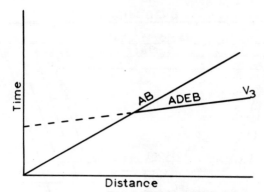

Fig. 13.11.   Plots of distance versus time for layers in the
             sub-bottom determined by seismic refraction
             shooting.

travel vertically upward to the hydrophone from the
stratum. Such a system works effectively under water
because the impedance match between the water and the
bottom sediments is quite good. Such a system would not
work effectively over land because of the poor impedance
match between air and land.

13.7.   *Underwater Communication.*

Since sound is the only form of energy that is propagated over appreciable distances under water, it was
natural to use it to communicate between submarines and
between submarines and ships on the surface. Since
World War II there has been considerable development
of equipment and technology for underwater swimmers.
Underwater telephones are an important part of that
equipment, because underwater swimmers must be able
to communicate with each other and with service boats

on the surface in order to avoid accidents and to report findings during their work. Because of the poor impedance match between air and water, the sound is picked up by means of a microphone, the signal is amplified and fed to an underwater transducer which serves as a projector. It is received by a hydrophone, amplified and fed to a projector in the air.

Oceanographers working on projects such as Sealab I and Sealab II need to communicate with each other and with the supporting ships on the surface. The underwater telephone is an important means for such communication.

When underwater swimmers work at great depths, they breathe an atmosphere of helium and oxygen to avoid breathing nitrogen at high pressure, which is the factor responsible for the "bends." When the swimmers breathe the helium-oxygen mixture, the quality of their voices is changed so that when they speak they sound like the "Donald Duck" of the cartoon movies. Because of the difficulty of understanding such speech, special electronic signal processing equipment has been developed, which converts such "Donald Duck"-like speech to more normal sounding speech, making it easier to understand the speaker.

As the level of activity in oceanography increases and underwater swimmers go to greater and greater depths, the importance of communicating by underwater sound will increase.

13.8. *Miscellaneous Applications of Underwater Sound.*

There are many miscellaneous applications of underwater sound and every year new ones are developed as new needs arise.

In positioning devices such as underwater cameras, which are to be located at a fixed distance above the bottom, a special underwater distance measuring system is

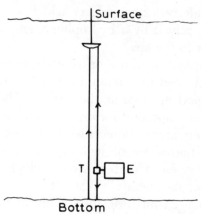

Fig. 13.12.   Method of positioning equipment above the bottom by use of acoustic signals.

used. Figure 13.12 shows the arrangement of this device. E is the equipment to be positioned, T is a transmitting transducer which transmits a pulse of sound toward the bottom and simultaneously transmits one toward the surface. The pulse transmitted toward the bottom is reflected from the bottom and arrives at the surface later than the pulse transmitted toward the surface by the amount of time it takes for the sound to travel from the projector to the bottom and back up to the level of the projector again. The time interval between arrival of the two pulses is a measure of the distance of the projector from the bottom. Since a sound pulse travels about five feet in a millisecond, the pulses will be separated by two milliseconds if the projector is positioned five feet above the bottom. The measurement of the distance from the projector to the bottom is independent of the depth below the surface.

Underwater sound navigation buoys are becoming quite important in connection with oil well drilling equipment. It is possible, by means of acoustic navigation buoys, to position a well drilling ship accurately enough to maintain a drill rod vertically over a well and even to reinsert a drill rod in the well hole after the drilling ship has had to leave the location because of a storm. The signals from the navigation buoys planted on the ocean bottom are received on hydrophones and processed by a computer, which controls thruster propellers on the ship to move it into position and maintain that position.

Oceanographers use buoys, planted below the surface of the ocean with recording equipment, to record various oceanographic parameters. These buoys must be recovered periodically to remove the records and to renew the batteries. A common method for release of the buoy is to have an acoustic receiving system on the buoy which will respond to a special coded signal and release the buoy from its anchor so that it will float to the surface. When batteries and recording material have been replaced, a new anchor is attached and the buoy is replaced.

With the availability of transistors it has been possible to develop small SoNARS that operate over short ranges at high frequencies. Such SoNARS have been produced with a pair of handles that can be grasped by an underwater swimmer and, as he swims, he can locate objects with the SoNAR that are beyond his range of vision.

14.

Utilization of Sound by Animals

14.1. *Introduction.*

Unless we give the matter some careful thought, we are apt not to realize the extent to which we and the other animals utilize sound. We are, of course, fully conscious of our use of sound in aural communication and in our enjoyment of music. We are conscious of birds singing and dogs barking, but many of the animal sounds which we accept as an extraneous factor in our environment are produced for some purpose.

14.2. *Utilization of Sound by Man.*

We are all conscious of our use of sound in conversation and music. However, we unconsciously use it in many other ways. In Section 7.6, the sound spectrograph was described as an instrument that can analyze vocal and other impulsive sounds. By use of the sound spectrograph, the voice of an individual can be identified by comparing a recording with a known record of the individual's voice. Our ears, with the analysis system in our brains, is an excellent sound analyzer with which we almost instantly analyze the voice sounds of our acquaintances and, even though we cannot see them, we identify

them by comparing the analysis with the record that we
have stored ready for instant recall. In the same way we
recognize the presence of people we know by the sound
of their footsteps.

In our normal environment, we are continuously exposed to sounds. Many of these sounds are associated
with equipment, such as ventilators, which maintains the
desired conditions in our environment. If any of this
equipment malfunctions, the sound produced by it
changes and we are soon aware of the change. We use
the sound produced by vehicles to warn us of their approach. An important use of sound is as a danger signal.

You can perform an interesting experiment by keeping
a record for an entire day of all of the times in which you
utilized sound for some purpose besides normal conversation. You will probably be surprised with the result of
such an experiment, because unless we consciously observe it we are not aware of the large amount of information we receive as sound and utilize in our daily lives.

People who have lost their eyesight are trained to improve their ability to use sound to compensate for their
loss of vision. A blind person can carry a small, high frequency sound generator, which does not disturb the people around him, but which he can use to detect obstacles
by the reflected sound. Even without such a generator,
a nearby obstacle changes the general sound field so that
the presence of the obstacle can be detected. This is one
means by which we sense the presence of an intruder in
a darkened room. The blind man's cane is used to tap
on the ground as he walks along and the reflections of the
sound of the taps of the cane are used as a method of
detecting the presence of obstacles.

14.3.   *Use of Sound by Animals as Danger and Courting Signals.*

Animals that go in groups have signals that they use to warn the other members of the group when they sense danger.

If a turkey hen is traveling with her brood, she will make the sound "tut" if she senses danger and all of the little turkeys will instantly hide and remain hidden until she gives the all-clear signal, "koa, koa, koa." They will then instantly reappear.

If one of a group of beavers working on a dam or playing in the water senses danger, he will slap his flat tail on the surface of the water and all of the group will immediately dive for safety.

If a cat comes into the back yard where birds are feeding, the birds will immediately give their danger signal, which is different for each species of bird but apparently is understood by all of the others.

A rattlesnake will warn a potential enemy that he will strike if the enemy comes closer by shaking the rattles on his tail. Apparently the rattlesnake would prefer not to strike but his warning says that he will if necessary.

Many birds and animals produce special sounds when they are courting. These sounds apparently play the same role as conversations of human lovers.

Many animals, such as birds, produce sounds that do not seem to play an important role in their lives, but often make them attractive to us. When a bird sits in a tree and sings, outside of the mating season, it may be that he is making the sounds because he feels good and knows that he can sing. Such singing by the bird may play the same role in his life as singing by a human animal. When birds are in the habit of feeding at a feeding board in the yard, they soon learn that the people in the house will put out more food if they will come to the board and make sounds.

There have been many recordings made of bird sounds by bird lovers. These recordings have been made primarily for the purpose of study of the bird sounds for use in learning to identify them by sound. Relatively little effort has been made to correlate the information on these recordings with the behavior of the birds. Such a study would help us to understand the use that the birds make of these sounds.

Studies have been made of the sounds produced by the fish called croakers. These studies have been correlated with the behavior of these fish and there is evidence that they establish an area which they claim and they produce the sound to warn others to stay out of the area.

As we learn more about the habits of animals, we will no doubt learn of more instances where the sounds produced by the animals perform some useful functions in their lives.

If you will go to a place away from the noises of the city on a summer evening and sit quietly and listen, you will hear a myriad of insect sounds. The insects have various mechanisms for making sounds, such as body structures which they can scrape together. It is reasonable to suppose that if nature provides these mechanisms for generating sound, they must be provided for a purpose. Probably one of the purposes is as a mating signal but there may be others. We can learn what these are only through studying the behavior of these insects and determining how their behavior correlates with their production sound.

Such experiments are not easy to perform because they are not meaningful unless they are carried out with the insects in their natural environment. This is one of the major problems in studying the utilization of sound by marine animals. If we work with them in aquariums where they can be readily observed, the environment is artificial and may have a considerable effect on their sound production and on their response to sounds. Studies on hearing acuity of fish must be carried out in such aquariums. We do not know how this may affect their response and there is a serious problem in determining what is the true sound level in such a confined volume whose linear dimensions are often small compared to a wavelength of the sound.

14.4. *Use of Sound for Navigation and Food Hunting.*

Bats fly and hunt their food at night. It is well known that they emit pulses of sound while they are flying. Studies of the habits of bats have shown that they use sound in the same manner as a SoNAR operator does. They navigate around obstacles by sensing the echoes from the obstacles and they locate and home in on the insects they capture for food. Studies on bats indicate that they have a SoNAR system equivalent to the more sophisticated systems used by Naval ships. The methods of processing the returned echo signals is not the same in all bats. Some bats produce a sound pulse that sweeps a range of frequency while others transmit a constant frequency signal. Tests carried out with blindfolded bats show that they can successfully navigate around wires stretched through a room.

The ability of bats to capture flying insects requires

greater sophistication in their SoNAR system than the ability to navigate around fixed barriers. The bat must be able to process the echo signal to determine the range and the rate of change of the bearing of the insect target. When we consider that the bat's entire signal processing system weighs less than one gram and it does a signal processing job at least equivalent to that performed by a sophisticated SoNAR or RaDAR system, which weighs many pounds, we are forced to view it with respect.

Some of the moths, on which the bats feed, have a sound receiving system that senses the homing pulses transmitted by the bats so that they can take evasive action and are thereby more difficult to capture.

The sound systems used by bats and *cetaceans* (dolphins and whales) are quite spectacular. As a result, they have received a great deal of attention and their characteristics are quite well known, although we do not have a complete understanding of their mechanism of operation. Much work needs to be done to gain more knowledge about the utilization of sound by other animals. This knowledge can be gained only through study of animal behavior by people skilled in that field who also have a good knowledge of acoustics.

15.

Careers in Acoustics

15.1. *Introduction.*

The science of sound is an interdisciplinary science. By this we mean that it is a science that cooperates with other sciences or fields of study. There are other interdisciplinary sciences but the science of sound is unique in that nearly every other science is, in some instances, involved with it.

Most scientists and engineers working on acoustic related problems were trained in some other field and have had to learn about acoustics by further study because the problems on which they were working involved sound in some way. Because of its interdisciplinary nature, no college or university has a department of acoustics, but some universities have recently instituted graduate programs with acoustics as a specialty. These programs usually involve taking an advanced degree in fields such as physics or engineering with a specialty in acoustics. The Pennsylvania State University has an arrangement in the graduate school whereby a student may earn a Master of Science or a PhD degree in Engineering Acoustics under an interdisciplinary committee composed of members of several cooperating departments. Under this arrangement, a student may have earned his Baccalaureate or even his

Masters degree in one of the engineering fields, physics, speech, biology or even psychology. He will then pursue his graduate work in acoustics as it is related to his undergraduate field of specialization. By means of these various kinds of graduate programs a student can receive training in acoustics as it relates to the cooperating field in which he has received his undergraduate training.

15.2. *Physics*.

Acoustics has traditionally been a branch of physics. The physical acoustics scientist is concerned with the fundamentals of vibrating systems and the sound emission from such systems. He generally is not interested in the applications of the science in the many other fields. However, many of the people now working in applied acoustics received their training as physicists and later joined one of the other fields. This, of course, involved self training or on-the-job training in the other field.

So much needs to be done in the fundamentals of acoustic science that it would be desirable for those physicists who have specialized in acoustics to make their contributions in physical acoustics.

15.3. *Architectural Engineering*.

Architects and architectural engineers are responsible for the design of buildings, auditoriums and concert halls. In order for these structures to serve their purpose it is necessary to design them so that noise from outside the building and noise and vibration generated by equipment in the building will not disturb the occupants. In the case of factories, it is essential to design the work areas so that the workers are protected as much as possible from excessive noise. Auditoriums must be designed so that speech intelligibility at all of the seats will be satisfactory and concert halls must be designed so that the performers will feel comfortable while performing and the music will be pleasing to the audience. School buildings must be designed so that the activities of one class will not be disturbing to the students in other classes.

The architectural acoustician needs to have a knowledge of building materials, methods of construction of buildings and building codes so that he can operate with the architects and architectural engineers to arrive at realistic designs that also meet the acoustic requirements. A student who studies architectural engineering as an undergraduate and then specializes in acoustics in graduate studies will be well qualified to perform this service.

As the requirements for the seating capacity of auditoriums and concert halls increases and as the requirements for low noise levels in working areas become more stringent, the demand for people adequately trained in this area will increase. There is also a need for people trained in this area to conduct research for building material manufacturers in order to develop better materials for sound isolation in buildings.

15.4. *Fluid Mechanics.*

As vehicles such as aircraft and ships travel through their respective fluids, vibrations are induced due to the flow of fluid over the vehicle bodies. These vibrations can cause damage to the vehicle structure. They can cause excessive noise within the vehicle and cause excessive noise to be radiated to the environment around the vehicle.

In addition to vibrations induced in an aircraft due to the air flow over it, there is an additional problem with modern aircraft due to the high velocity gas flow in jet engines. The major airports are near to populated areas and the problem of jet aircraft noise is already critical and will become more serious as aircraft speeds increase.

Vibrations are produced in ships due to flow of water over surfaces and the flow over openings to cavities in the hulls often produces cavity resonances.

Propeller blades, fins and rudders on vehicles running in the water may produce noise by singing which is caused by the shedding of eddies in the flow over the edges of the surfaces. Cavitation, which is a very serious noise produced by vehicles running in the water, results from a flow condition where the pressure is reduced sufficiently to cause bubbles of water vapor to form and then collapse when they move to a higher pressure region.

Engineers who design ships and aircraft must have knowledge of both acoustics and fluid mechanics to apply the latest knowledge in these fields to the design of these structures. There is also a great need for engineers and scientists in these fields to carry out research programs to increase our fund of knowledge in these fields.

15.5. *Electrical Engineering.*

Communication engineering is a branch of electrical engineering and acoustics. The early designers of communication equipment and recording and reproducing equipment were not adequately trained in acoustics, and therefore did not understand all of the factors important in sound reproduction. For this reason, the early "high fidelity" sound equipment was not satisfactory to critical listeners. The broadcast and phonograph industries employ sound engineers, who are electrical engineers with a knowledge of acoustics, to design sound systems. Such engineers must have a good knowledge of electrical circuit design, a knowledge of electroacoustic transducers and a knowledge of the characteristics of the human hearing mechanism in order to successfully design the high-quality sound systems used in the modern communications industry.

15.6.   *Mechanical Engineering.*

A mechanical engineer who designs machines for use in industry is normally concerned with the machine efficiently doing its job. We now realize that high noise levels in the factory are a hazard to the hearing of the workers. The noise level generated by a machine is therefore a factor that must be considered in its design. The engineer, therefore, needs to have a knowledge of acoustics in order to design machines which have the lowest possible noise levels. Since many components such as motors and gear trains must be purchased from other suppliers, the engineer must know how to write the acoustic specifications for these components and he must know how to measure their noise output to determine if the required specifications have been met.

There are not many mechanical engineers with the necessary training in acoustics. This deficiency is met in some instances by the industries sending their engineers to universities to take short courses in acoustics. The ideal arrangement would be for the young engineer to take an additional year after receiving his degree in engineering to specialize in acoustics at one of the universities offering such a specialty in their graduate program.

15.7. *Biology.*

Biologists now realize the importance of sound in the animal world. In order to learn how animals utilize sound it is necessary to measure the characteristics of sounds emitted by the animals and correlate this information with the other behavior of the animal, and it is also necessary to study the way in which the animals respond to sound.

In order to make such studies, it is necessary for the observer to have a good knowledge of animal behavior as well as acoustics. In studying the animal's reaction to sound it is usually necessary to train it to do certain things when it perceives the sound and it is important to provide an environment which is not so artificial that the reaction of the animal will be unduly affected by it.

Most experiments with animals must be carried out in restricted areas, which pose a difficult problem for acoustic measurements. For example: when the hearing acuity of fish is being studied, it is not a simple matter to determine the sound level in the area occupied by the fish in the relatively small aquariums that must be used for such experiments.

This is an important and interesting field of research, but to obtain meaningful results the experimenter must have a thorough knowledge of animal behavior and a thorough knowledge of acoustic principles.

15.8. *Community Planning.*

We are beginning to realize the importance of noise in our environment. Important sources of noise in the community are traffic, airports, factories, music halls and children's playgrounds.

The community planner must have a knowledge of the kinds of noise and the levels of noise produced by the various kinds of noise sources. He must also know how to incorporate trees and shrubs and other means of sound absorption to minimize the effects of noise. He must also know which kinds of activity should be barred from residential communities. He must know how to write specifications for the noise levels from activities which are proposed for the community and he must know how to make the necessary noise measurements to determine if these specifications are being met.

We are quite conscious of the problems of pollution. Noise is one form of pollution which has been increasing at an alarming rate. Some experts claim that the noise level has been increasing at a rate of three dB per year. The community planners of the future will be responsible for designing communities in which this trend can be reversed.

15.9. *Public Health.*

It is only recently that the state legislatures and the congress have recognized that noise is a public health problem. Many of the state legislatures and the congress have passed laws for the protection of workers from hearing loss due to excessive noise in industry. The public health services are responsible for the enforcement of the standards set up by the legislative bodies and they must also work with the industrial health groups in the various industries in order to assist them in determining where noise hazards exist.

Management in industry is generally interested on cooperating with the public health services, since workers who have suffered hearing loss can draw workman's compensation.

It is necessary to have many more people than are now available who are skilled in the field of industrial noise. Some of these people will be employed by industry and some in the public service. They are concerned with the measurement of noise levels at the working position in factories and in the measurement of temporary and permanent hearing threshold shifts in the workers. They are also responsible for studying the data that is being accumulated on hearing threshold shifts to determine the necessary length and frequency of rest periods for workers under the various industrial noise conditions in order to insure adequate protection for their hearing. When ear protectors must be used it is necessary to have people who have the necessary skills to specify the proper type of ear protector and to properly fit the ear protectors to the individual workers.

Some universities have departments of speech and hearing pathology. A student trained in acoustics and hearing pathology would be well equipped for a career in industrial noise. There is now a shortage of people with adequate training in this important field and the shortage will become more critical in the future.

15.10. *Oceanography*.

Oceanography is a rapidly growing science, and acoustics is one of its important tools. Acoustics is used in basic oceanographic research and in applied oceanography. It is used in the precision depth sounder to map ocean bottom contours and it is one of the important means of studying the deep scattering layers. It is used in fishermans SoNARS to locate fish for the fishing industry and to study the habits of fish.

An oceanography student should study acoustics, particularly underwater acoustics, in order to be adequately prepared to make the most effective use of this important tool whether the work is in basic or applied oceanography.

The oceanographer utilizing acoustics must have a thorough knowledge of underwater acoustic transducers. He must have a good understanding of sound propagation in the ocean and the sound reflecting characteristics of marine animals.

15.11. Marine Geology.

Underwater acoustics is an invaluable tool for the marine geologist. He must be familiar with the use of explosive sound sources and the underwater spark for use in studies of the layers in the sub-bottom. He must also have a knowledge of electroacoustic transducers which are used to receive the reflected acoustic signals.

Acoustic methods are used in basic marine geology research and they are also used in surveys to locate the various sediment and rock layers in regions where undersea oil wells are drilled and where other mineral resources are to be mined. An interesting application of acoustic methods was the survey of the sediment and rock layers in the region where the proposed tunnel is to be constructed under the English Channel.

Supplementary Reading

The following books are recommended for additional reading. Some do not require appreciable mathematical background and some require considerable mathematics. The first book listed is out of print but copies may be found in some libraries. It is listed here because it contains descriptions of many sound experiments that can be performed with simple apparatus.

1. *On Sound,* John Tyndall, D. Appleton Co., New York (1873).
2. *Music, Acoustics and Architecture,* Leo Beranek, John Wiley and Sons, New York (1962).
3. *The Physics of Music,* Alexander Wood, Dover Publications, New York (1944).
4. *Music Physics and Engineering,* Harry F. Olson, Dover Publications, New York (1967).
5. *Acoustics and Vibrational Physics,* R. W. B. Stephens and A. E. Bate, Edward Arnold, Ltd., London (1966).
6. *Underwater Acoustics Instrumentation,* Vernon M. Albers, Instrument Society of America, Pittsburgh, Pa. (1969).
7. *Handbook of Noise Measurements,* Arnold G. P. Peterson and Ervin E. Gross, Jr., General Radio Co., West Concord, Mass. (1963).

8. *High Quality Sound Reproduction,* James Moir, The Macmillan Co., New York (1958).
9. *Fundamentals of Acoustics,* Lawrence E. Kinsler and Austin R. Frey, John Wiley and Sons, New York (1962).
10. *Acoustics,* Leo L. Beranek, McGraw Hill, New York (1954).
11. *Underwater Acoustics Handbook II,* Vernon M. Albers, The Pennsylvania State University Press, University Park, Pennsylvania (1965).
12. *The Theory of Sound,* Lord Rayleigh, Dover Publications, New York (1945).

Index

Absorption coefficient, 181, 183
Accelerometer, 77
Acoustic, axis, 197
  Cavitation, 195
  Impedance, 31, 35, 54, 61, 84, 85, 182, 193, 194, 212
  Power, 195
Acoustics, 9, 37, 224
Aeolian, 157
Aircraft noise, 113, 123
Ambient noise, 73
Ambient pressure, 86
Ammonium dihydrogen phosphate, 83
Amplification, 188
Amplifier, 96, 97
  Phonograph, 131
Analysis, 143
  Narrow band, 104
Antinode, 169
Antisubmarine warfare, 192
Anvil, 54
Archeology, 193
Architectural engineering, 225
Armature, 88
Array, 93
  Directional, 94
  Line, 93
  Transducer, 92, 194
Astrodome, 190
Attenuation, 29, 68, 198
  Coefficient, 68
Attenuator, 98
Audiogram, 58, 112
Audiometry, 112
Auditorium, 183, 226

Banjo, 163
Bar, 46
Barium titanate, 78
Basilar membrane, 55
Bathythermograph, 201
Bats, 222
Bearing, 196
Beats, 103
Békésy, 56
Bel, 46
Bells, 178
Bimorph transducer, 85
Biology, 229
Biot, 10
Bone conduction, 121
Boner, 189
Booms, supersonic aircraft, 123
Borelli, 10
Bottom contours, 209
Bottom topography, 208
Bowed strings, 163
Bridge, 20 166

Cardioid microphone, 82
Cassini, 10
Cavitation, 227
  Acoustic, 195
Cello, 169
Cents, 152
Ceramic transducer, 83, 84, 85, 86
Cetaceans, 223
Chladni, 12, 21
Chromatic tetrachord, 155
Clarinet, 176
Clouds, 186
Cochlea, 54

Colladon, 10
Communication, aural, 218
  Engineering, 227
  Underwater, 214
Community noise, 123, 230
Compressional wave, 36
Concert hall, 182, 186, 190, 226
Concha, 53
Condenser microphone, 90, 95, 98
Cornet, 176
Crystal transducer, 79, 83, 85
Curie, 78
Cycle, 18
Cymbal, 178

Damping, 23
DBs, 46
Decibel, 38, 39, 46, 60
Deep sound channel, 205
Detector, pressure, 36
  Velocity, 36
Diapason pipe, 173
Diaphragm, 81, 86, 90
Diatonic tetrachord, 155
Difference tones, 11
Directional arrays, 94
Directivity, 194
  Factor, 92
  Index, 93, 94
  Transducer, 92
Distortion, 139, 189
  Frequency, 139
  Harmonic, 140
  Intermodulation, 141
  Tracing, 133
Divergence, 60, 69, 198, 205
Downward refraction, 204
Drum, 176
Dulcimer, 163
Dynamic range, 110

Ear, 9, 36, 52, 56, 58, 132, 145, 218
  Drum, 53
  Fish, 207
  Inner, 53
  Middle, 53

Muffs, 121
Outer, 53
Plugs, 121
Protection, 120
Protectors, 58, 128
Echoes, 10, 185
Echo ranging, 193
Eddy current, 88
Edison phonograph, 125
Electrical engineering, 227
Electroacoustic transducer, 195
Electrodynamic transducer, 78, 86, 87
Electronic organs, 150, 179
Electrostatic transducer, 78, 80
Electrostrictive transducer, 78
Embouchure, 176
Expander bar transducer, 84
Expander plate transducer, 84
Explosive sources, 94
External noise, 73

Factory noise, 123
Fathometer, 193, 206, 208
Feedback, negative, 189
  Positive, 189
Filter, 74, 96
  Characteristics, 74
  Center frequency, 100
  One-half octave, 74, 100
  One octave, 74
  One-third octave, 74
Flue pipe, 172
Fluid mechanics, 226
Flute, 176
French horn, 176
Frequency, 10, 19
  Analyzer, 101
  Band, 23
  Distortion, 139
Fundamental, 10, 20

Galileo, 10
Gaussian, distribution, 72
  Random noise, 72
Glottal, attack, 50
  Stop, 50

Greek, Aeolian, 157
  Scales, 155
  Tetrachord, 163
  Theatres, 180
Guitar, 163

Hairpin stack, 88
Hammer, 54
Harmonic, 50
  Distortion, 140
Harmonics, 160
Harmonious, 149
Harmony, 154
Harp, 163
Harpsichord, 163
Hayes, 208
Hearing, damage, 58
  Fish, 222
  Impairment, 118
  Threshold, 112, 118
  Threshold shift, 114
Heart beat recording, 110
Helicotrema, 56
Helmholz, 11, 33, 56, 149
Hertz, 37
High-fidelity, 139
Homophonic music, 153
Huygens, 10
Hydrophone, 77, 85, 86, 95

Impedance, acoustic, 31, 35
  Match, 126
  Transformation, 54
Impulse noise, 119
Industrial noise control, 120
Infrasonic, 37
Inner ear, 53, 55
Input noise, 96
Intensity, 11, 33, 195
  Level, 39
  Sound, 34
Interference, 10
Intermittent noise, 116
Intermodulation distortion, 141
Internal waves, 200
Intonation, 147
Isothermal layer, 200

Isovelocity, 70

Joule, 34

Keinetic energy, 29
Kettle drum, 177
Koenig, 12
Kundt, 22

Lagrange, 10
Laminated stack, 88
Laminations, 88
Laplace, 11
Larnyx, 50
Lateral line, 207
Layered medium, 70
Lead zirconate titanate, 78
Level, 39, 49
  Noise, 75
  Signal, 75
Line array, 93
Lithium sulphate, 84
Logarithm, 46
Loudness, 39, 57, 145
Loud speaker, 79, 92
Lydian, 155

Magnetic recording, 135
  Tape recording, 136
Magnetostrictive, 88
  Transducer, 78, 80, 88
Major scale, 152
Marine, animals, 206
  Geology, 193, 211, 232
Masking, 113
Meatus, 53
Mechanical, engineering, 228
  Impedance, 85
Medium, layered, 70
Mel, 146
Melon, 208
Mersenne, 10
Microbar, 46
Microphone, 12, 35, 38, 43, 77,
  80, 86, 95
  Cardioid, 82
  Condenser, 90, 95, 98

Pressure actuated, 80
Pressure-gradient actuated, 80
Middle ear, 53, 55
Minor scale, 152
Modes, 19, 20, 21
    Radial, 86
Modulation, 142
Monochord, 10
Music, 9, 144, 218
Musical, intervals, 9, 150
    Scales, 153
    Sound, 144
Music box, 158

Navigation buoys, 217
Negative feedback, 189
Negative feedback loop, 190
Network, weighting, 99
Newton, 10, 35
Nodes, 21, 23, 159
Noise, 12, 72, 112, 192
    Aircraft, 113, 123
    Addition of, 75
    Ambient, 73
    Community, 123
    Environment, 115
    Excessive, 114
    Exposure, acceptable, 117
    External, 73
    Factory, 123
    Gaussian random, 72
    Hazard, 98, 112, 115
    Hazard, legal aspects, 122
    Impulse, 119
    Industrial, 120
    Input, 96
    Intermittent, 116
    Jet aircraft, 227
    Level, 75
    Random, 72
    Reduction, 76, 120
    Self, 73
    Thermal, 96
    Traffic, 123
    White, 72
Non reciprocal transducers, 94, 196

Oceanography, 215, 231
Octave, 10, 37, 151
    Band width filter, 74
Ohm, 12
Omnidirectional, 92, 97
Optical recording, 133
Organ, 171
    Electronic, 179
Outer ear, 53
Oval window, 55

Particle velocity, 35, 38, 81
    Detector, 36
Party effect, 114
Pattern, 82, 93
    Sum and difference, 196
Permanent threshold shift, 115
Permendur, 89
Phase, 28, 30
Phon, 57
    Scale, 58
Phonautograph, 12
Phonograph amplifier, 131
    Needle, 128
    Stylus, 130
    Transducer, 87, 129
Physics, 225
    Of music, 144
Pianoforte, 165
Picard, 10
Piezoelectric transducer, 78, 83
Pipes, 170
Pipe organ, 175
Pitch, 145
    Pipes, 148
    Recognition, 147
Plectrum, 164
Plucked strings, 163
Polarized, 78, 80, 88
Polyphony, 154
Porpoises, 208
Positive feedback, 189
Potential energy, 28
Preamplifier, 98
Precision depth sounder, 209
Predator, 208
Pressure-gradient microphone, 81

Projector, 77
Propagation of sound, 24
Public health, 123, 230
Pulsed signals, 195
Pythagoras, 9, 150

Q, 24
Quality, 148

Radial mode, 86
Random noise, 72
Ray, 61, 67
Rayleigh, 11, 13, 64
Reciprocal transducer, 77, 86, 90, 195
Recorder, 176
 Sound level, 101
Recording, heart beat, 110
 Magnetic, 135
 Optical, 133
 Sound, 125
Reed pipe, 174
Reference sound pressure, 41
Reflection, 30, 61, 182
Reflector, 64, 186
Refraction, 65, 202
 Downward, 204
 Upward, 204
Resonance, 24, 92
Resonator, 11, 33, 50
Response pattern, 82
Reverberation, 10, 181, 185
 Time, 181, 183
Ring stack, 88
Rochelle salt, 83
Rock'n'roll, 9, 59
Römer, 10
Round window, 55

Sabine, 183
Salinity, 199
Salinometer, 201
Saunders, 53
Saxophone, 176
Scales, 152
 Greek, 155
Scattering layers, 206

Searchlight mode, 196
Seawater sound attenuation in, 199
Seismic refraction shooting, 212
 Waves, 211
Self noise, 73
Semitone, 152
Sensitivity, 43, 45, 92
Shadow zone, 203, 204
Shear plate transducer, 84
Shift of hearing threshold, 114
Signal, level, 75
 Pulsed, 195
Simple harmonic motion, 18
Snare drum, 177
Snell, 66
Snell's law, 66, 69, 202
SoNAR, 192, 208, 210, 217, 222
Sone scale, 58
Sonic, 37
Sound, 9, 36, 219, 224
 Analysis, 143
 Attenuation, 199
 Channel, 205
 Channel axis, 206
 Insulation, 181
 Intensity, 34
 Intensity level, 49
 Level, 43, 99
 Level meter, 98
 Particle velocity, 193
 Power level, 49
 Pressure, 35, 38, 96, 193
 Pressure level, 39, 49
 Pressure measurement, 96
 Projector, 85
 Propagation, 24, 67, 69, 192, 198
 Recording, 125
 Reinforcement, 93, 188
 Reproduction, 125, 228
 Spectrograph, 106, 218
 Velocity profile, 201
 Velocity, seawater, 199, 200
Sounding board, 162
Sounds, bird, 221
 Fish, 221

Insect, 221
  Musical, 9, 144
  Vocal, 50
Sources, explosive, 94
Spectrograph, sound, 106
Spectrum analysis, 101
  Narrow band, 102, 104
Speech, 50
Spherical mirror, 64
Stack, hairpin, 88
  Laminated, 88
  Ring, 88
Standing wave, 30, 33, 159
Stereophonic effect, 132
Stirrup, 54
Strings, vibrating, 162
Sturm, 10
Subbottom, 211
  Stratum, 213
Subjective tones, 147
Submarine, 192
Summation tones, 11
Supersonic aircraft booms, 123
Surface layer, 200
Swim bladders, 207

Tambourine, 176
Temperature gradient, negative, 203
  Positive, 202
Temperature inversion, 67
Temporary threshold shift, 115
Termination, 30
Tetrachord, 154
Theater, Greek, 180
Thermal noise, 96
Thermocline, 200
Threshold of hearing, 112
  Shift, permanent, 115
  Shift, temporary, 115
Timbre, 148
Tone, 152
Traffic noise, 123
Transducer, 77, 192, 193
  Arrays, 92, 93
  Bimorph, 85

Ceramic, 85
Ceramic ring, 86
Crystal, 85
Directivity, 92
Directivity factor, 92
Directivity index, 93, 94
Electrodynamic, 86, 87
Electrostatic, 78, 80
Electrostrictive, 78
Elements, 78
Expander bar, 84
Expander plate, 84
Line array, 93
Magnetostrictive, 78, 80, 88
Non reciprocal, 94, 196
Omnidirectional, 97
Phonograph, 87, 129
Reciprocal, 77, 86, 90
Ring type, 194
Shear plate, 84
Underwater sound, 88
Wire recorder, 135
Transients, 143
  Response, 143
Transmission loss, 182
Trombone, 176
Tuba, 176
Tuning fork, 147, 161
Tyndall, 11

Ultrasonic, 37
Underwater, acoustics, 192
  Cameras, 216
  Positioning equipment, 216
  Sound detection, 193
  Sound generation, 193
  Sound navigation buoys, 217
  Sound transducer, 88
  Swimmers, 214
  Telephone, 214
Upward refraction, 204

Variable, area recording, 133
  Density recording, 133
Velocimeter, 201
Velocity gradient, 205

Of sound, 10
Vibrating, rod, 19
  String, 19
  System, 19
Vibration, of air columns, 169
  Of plates, 21
Vibrator, mass spring, 17
Viola, 169
Viol family, 169
Violin, 167
Viviani, 10
Vocal chords, 50
  Folds, 50
  Sounds, 50
Voice prints, 110

Watt, 34
Wave, 24
  Compressional, 27, 36
  Fronts, 33, 62
  Standing, 30, 33
Wavelength, 28
Weighting network, 99
Weston, 207
Whales, 208
Wheatstone, 12
White noise, 72
Willow whistle, 173
Wire recorder transducer, 135

Xylophone, 160